Functional Mapping of the Brain in Vascular Disorders

Edited by
Wolf-Dieter Heiss

With 30 Figures and 14 Tables

Springer-Verlag
Berlin Heidelberg New York Tokyo

Professor Dr. med. WOLF-DIETER HEISS
Max-Planck-Institut für neurologische Forschung
Abteilung für allgemeine Neurologie und Neurologische Klinik
Krankenhaus Merheim, Ostmerheimer Straße 200
5000 Köln 91, FRG

ISBN-13:978-3-540-15801-1 e-ISBN-13:978-3-642-70720-9
DOI: 10.1007/978-3-642-70720-9

2125/3130-543210

Preface

This book contains the contributions to the symposium "Functional Mapping of the Brain in Vascular Disorders", held at the Thirteenth World Congress of Neurology, September 1–6, 1985 in Hamburg, FRG. I have to thank the contributors to this symposium for submitting their manuscripts long before the congress so that the printed proceedings could be distributed to the audience. I hope that this will enable the participants in this symposium not only to recall the vivid presentation of the lectures and the highlights of the discussions, but also to widen their knowlegde of the topics dealt with during the symposium by rereading the chapters on the various issues. I would also like to express my thanks to the company UCB, Kerpen, FRG, who supported the symposium and the printing of these proceedings.

Cologne, July 1985 W.-D. Heiss

Contents

Contributors

Beil, C., Dr., Max-Planck-Institut für neurologische Forschung, Abteilung für allgemeine Neurologie, und Neurologische Klinik, Krankenhaus Merheim, Ostmerheimer Straße 200, 5000 Köln 91, FRG

Cho, Z. H., Dr., Columbia University College of Physicians and Surgeons, The Neurological Institute, 710 West 168th Street, New York, NY 10032, USA

Diksic, M., Dr., Montreal Neurological Institute and Hospital, Neuroimaging Research, McGill University, 3801 University Street, Montreal Quebec, Canada H3A 2B4

Evans, A. C., Dr., Montreal Neurological Institute and Hospital, Neuroimaging Research, McGill University, 3801 University Street, Montreal Quebec, Canada H3A 2B4

Feindel, W. H., Prof. Dr., Montreal Neurological Institute and Hospital, Neuroimaging Research, McGill University, 3801 University Street, Montreal Quebec, Canada H3A 2B4

Frackowiak, R. S. J., Dr., Medical Research Council, Cyclotron Unit and Department of Neurology, Hammersmith Hospital, Ducane Road, London W12 OHS, England

Gibbs, J. M., Dr., National Hospital, Queen Square, London WC1E 3BG, England

Hakim, A. M., Dr., Montreal Neurological Institute and Hospital, Neuroimaging Research, McGill University, 3801 University Street, Montreal Quebec, Canada H3A 2B4

Heiss, W.-D., Prof. Dr., Max-Planck-Institut für neurologische Forschung, Abteilung für allgemeine Neurologie, und Neurologische Klinik, Krankenhaus Merheim, Ostmerheimer Straße 200, 5000 Köln 91, FRG

Herholz, K., Dr., Max-Planck-Institut für neurologische Forschung, Abteilung für allgemeine Neurologie, und Neurologische Klinik, Krankenhaus Merheim, Ostmerheimer Straße 200, 5000 Köln 91, FRG

Herold, S., Dr., Medical Research Council, Cyclotron Unit and Department of Neurology, Hammersmith Hospital, Ducane Road, London W12 OHS, England

Hilal, S. K., Prof. Dr., Columbia University, College of Physicians and Surgeons, The Neurological Institute, 710 West 168th Street, New York, N. Y., 10032, USA

Jones, T., Prof. Dr., Medical Research Council, Cyclotron Unit and Department of Neurology, Hammersmith Hospital, Ducane Road, London W12 OHS, England

Kogure, K., Prof. Dr., Tohoku University, Department of Neurology, Institute of Brain Diseases, 1–1 Seiryo-machi, Sendai 980, Japan

Kohno, H., Dr., Central Research Laboratory Hitachi Ltd. p. o. Box 2, Kokubunji, Tokyo 185, Japan

Lenzi, G. L., Dr., III. Clinica Neurologia, Viale dell' University 30, 00185 Roma, Italy

Matsui, S., Dr., Central Research Laboratory, Hitachi Ltd. p. o. Box 2, Kokubunji, Tokyo 185, Japan

Maudsley, A. A., Dr., Columbia University, College of Physicians and Surgeons, The Neurological Institute, 710 West 168th Street, New York, NY 10032, USA

Meyer, E., Dr., Montreal Neurological Institute and Hospital, Neuroimaging Research, McGill University, 3801 University Street, Montreal Quebec, Canada H3A 2B4

Mun, S. K., Dr., Columbia University, College of Physicians and Surgeons, The Neurological Institute, 710 West 168th Street, New York, NY 10032, USA

Ohtomo, H., Dr., Tohoku University, Department of Neurology, Institute of Brain Diseases, 1–1 Seiryo-machi, Sendai 980, Japan

Pawlik, G., Dr., Max-Planck-Institut für neurologische Forschung, Abteilung für allgemeine Neurologie, und Neurologische Klinik, Krankenhaus Merheim, Ostmerheimer Straße 200, 5000 Köln 91, FRG

Pokrupa, R. P., Dr., Montreal Neurological Institute and Hospital, Neuroimaging Research, McGill University, 3801 University Street, Montreal Quebec, Canada H3A 2B4

Ra, J. B., Dr., Columbia University, College of Physicians and Surgeons, The Neurological Institute, 710 West 168th Street, New York, NY 10032, USA

Roschmann, P., Dr., Philips Medical Systems, Inc., 710 Bridgeport Avenue, Shelton, CT 06484, USA

Simon, H. E., Dr., Philips Medical Systems, Inc., 710 Bridgeport Avenue, Shelton, CT 06484, USA

Syrota, A., Prof. Dr., Commissariat à l'Energie Atomique, Dépàrtment de Biologie, Service Hospitalier Frédéric Joliot, Hôspital d'Orsay, 91406 Orsay, France

Ter-Pogossian, M. M., Prof. Dr., Mallinckrodt Institute of Radiology, Washington University, School of Medicine, 510 South Kingshighway, St. Louis, Missouri 63110, USA

Thompson, C. J., Dr., Montreal Neurological Institute and Hospital, Neuroimaging Research, McGill University, 3801 University Street, Montreal Quebec, Canada H3A 2B4

Tyler, J., Dr., Montreal Neurological Institute and Hospital, Neuroimaging Research, McGill University, 3801 University Street, Montreal Quebec, Canada H3A 2B4

Wagner, R., Dr., Max-Planck-Institut für neurologische Forschung, Abteilung für allgemeine Neurologie, und Neurologische Klinik, Krankenhaus Merheim, Ostmerheimer Straße 200, 5000 Köln 91, FRG

Wienhard, K., PD Dr., Max-Planck-Institut für neurologische Forschung, Abteilung für allgemeine Neurologie, und Neurologische Klinik, Krankenhaus Merheim, Ostmerheimer Straße 200, 5000 Köln 91, FRG

Wise, R. J. S., Dr., National Hospital, Queen Square, London WC1E 3BG, England

Wittekoek, S., Dr., Philips Medical Systems, Inc., 710 Bridgeport Avenue, Shelton, CT 06484, USA

Worsley, K., Dr., Montreal Neurological Institute and Hospital, Neuroimaging Research, McGill University, 3801 University Street, Montreal Quebec, Canada H3A 2B4

Yamamoto, Y. L., Prof. Dr., Montreal Neurological Institute and Hospital, Neuroimaging Research, McGill University, 3801 University Street, Montreal Quebec, Canada H3A 2B4

The Purpose of Functional Mapping in Focal Cerebral Ischemia

W.-D. Heiss

Experimental results obtained during the last decade [3, 5, 6] have demonstrated that cerebral cell damage following focal ischemia is not caused by a single event, but develops as the result of a cascade of pathogenetic steps. The primary event is the regional fall in cerebral blood flow (CBF) below a critical threshold, which differs for the maintenance of function and morphological integrity of the nerve cells [1]. Insufficient blood and oxygen supply causes alterations in metabolism, especially glycolysis, and depletion of primary (ATP, PCr) and secondary (glucose) energy reserves; leads to changes in the electrophysiological properties of the nerve cells; and finally causes the depolarization of membranes by excessive changes in ion homeostasis. The fate of a cell is determined not only by the severity of the ischemic event but also by its duration, which even at the initial stage allows a limited period of time during which complete recovery is possible if reperfusion above a critical value can be achieved.

The abnormal ion homeostasis resulting from the disturbed membrane function — an increase in extracellular potassium concentration and an influx of sodium and calcium ions into the cell — triggers a series of pathogenetic mechanisms, including the liberation of free fatty acids. With the availability of oxygen during reperfusion, it also gives rise to the production of various substrates which further damage the cell membranes, such as leukotrienes, prostaglandins, and endoperoxides. Continuing glycolysis during the ischemic period leads to the excessive production of lactate and a fall in pH; the developing cellular acidosis constitutes another factor which aggravates tissue damage. All these chemical mechanisms, which are secondary to ischemia, have a direct effect on cell metabolism and membrane function, but they also disturb the blood-brain barrier (BBB) and, as a consequence, brain edema develops. The increase in extracellular water further impairs the supply of oxygen and metabolic substrates to the cells, and the clearance of potentially harmful metabolic products from the tissue.

The data shedding light on the development of ischemic cell damage and the pathogenetic chemical mechanisms involved were obtained mostly by tissue sampling techniques or autoradiographic methods in animal experiments. Therefore, the results obtained are usually descriptive of one step in the pathogenetic cascade or one point in the time course, but they are of only limited conclusiveness for the temporal development of a sequence of pathogenetic steps. Multiparametric imaging of various variables with these techniques is also restricted to one time point: such experiments demonstrate concomitant alterations of the investigated variables (e.g., EEG, evoked potentials, CBF, glucose content and utilization, ATP content, protein synthesis, tissue pH, potassium concentration [2]) but do not permit any prediction of the further development of these changes, of their impact on the final outcome or of effects of treatment on the pathologic alterations. Usually, these experiments cover only the acute stage after focal ischemia; data on changes in later stages in the course after ischemia are still scarce. Some information on the development of altered variables in the course of focal ischemia and on the final functional and morphological outcome can be gained by serial recordings in chronic experiments [4].

In order to transfer the results from animal experiments to the clinical setting and to apply the knowledge acquired to the treatment of stroke patients, methods which permit repeated study of flow and metabolism are necessary. The imaging techniques used for the detection of morphological changes — especially x-ray computed tomography (CT) and magnetic resonance imaging of protons (MRI) — permit the demonstration of infarcts, of perifocal edema, and of disruption of the BBB, but pathophysiologic alterations leading to irreversible tissue damage cannot be visualized. By new techniques for the imaging of various parameters connected to function, especially positron emission tomography (PET) and MRI of ^{23}Na and ^{31}P, these changes can be visualized and quantified, and these studies repeated because of their noninvasiveness. On the basis of the observed changes of various variables, therapeutic measures aimed at the normalization of pathogenetic mechanisms can be applied, their effect on the target variables quantified, and their influence on the clinical course evaluated. Under these circumstances the lower spatial resolution of the techniques applicable in humans in comparison to the autoradiographic methods in animals is more than counterbalanced by the repeatability of the atraumatic clinical investigations.

It is the purpose of this special symposium and of this book to demonstrate the value of the mapping and quantifying of various functional parameters for our understanding of the pathogenesis of focal cerebral ischemia and ischemic damage in humans and for the development of therapeutic strategies in stroke victims.

References

1. Heiss WD (1983) Progress in cerebrovascular disease: flow thresholds for functional and morphological damage of brain tissue. Stroke 14:329–331
2. Hossmann KA, Mies G, Paschen W, Csiba L, Bodsch W, Rapin JR, Poncin-Lafitte M Le, Takahashi K (1985) Multiparametric imaging of blood flow and metabolism after middle cerebral artery occlusion in cats. J Cereb Blood Flow Metab 5:97–107
3. Raichle ME (1983) The pathophysiology of brain ischemia. Ann Neurol 13:2–10
4. Sato M, Paschen W, Pawlik G, Heiss WD (1984) Neurologic deficit and cerebral ATP depletion after temporary focal ischemia in cats. J Cereb Blood Flow Metab 4:173–177
5. Siesjö BK (1981) Cell damage in the brain: a speculative synthesis. J Cereb Blood Flow Metab 1:155–185
6. Siesjö BK (1984) Cerebral circulation and metabolism. J Neurosurg 60:883–908

Positron Emission Tomography Versus Nuclear Magnetic Resonance Imaging?

M. M. Ter-Pogossian

In the course of the last decade, positron emission tomography (PET) and nuclear magnetic resonance (NMR) imaging have evolved from promising laboratory experiments to mature medical imaging procedures with well-recognized places in the biomedical armamentarium. It is now well established that in many instances each of these modalities provides information about the patient of a nature which has not so far been obtainable by any other means. In the recent past, the question has been raised by scientific investigators whether there was any appreciable overlap between the information supplied by PET and that obtainable from NMR imaging and, if such an overlap does in fact exist, whether in such areas of overlap one of the modalities would not displace the other if it were in any way superior.

It is the purpose of this study to analyze the possible areas of overlap between PET and NMR imaging and to assess the relative strengths of these two modalities if such areas of overlap are identified. The following analysis is based on our present knowledge of the physical capabilities of PET and NMR imaging, and also on extrapolations about future improvements which can reasonably be made from our present understanding of the capabilities and limitations of these imaging approaches, without, however, postulating the development of new technologies which cannot be directly extrapolated from our understanding of the physical principles governing PET and NMR imaging. From the practical standpoint, this means an evaluation of these modalities as they stand now and for their future for the next 10 years.

It is interesting to note that there is a degree of parallelism between the histories of PET and NMR imaging. The seeds of the underlying concepts for these two technologies "germinated" during the 1940s. For PET, it was the recognition that a small number of short-lived positron-emitting radio-nuclides, particularly ^{11}C, ^{13}N, and ^{15}O (the "physiological" radionuclides), possess chemical and physical properties which render them particularly attractive for the in vivo investigation of some of the biochemical processes essential to life. For NMR imaging, it was the observation that the magnetic moments exhibited by some nuclides could lead to their identification in a sample by nondestructive means, and also that the measure of the NMR signal on a frequency scale could yield precious information about the chemical and physical states of the molecules containing the NMR-signal-producing nuclides. In the decades which followed, the above properties were extensively utilized for NMR in the study of chemical processes and, more specifically, of biochemistry. For PET, these properties were exploited in the use of "physiological" positron-emitting radionuclides for the in vivo and noninvasive study of some parameters unobtainable by other means, such as blood flow and metabolism of organs, particularly in the brain and heart. In the early 1970s, spurred in all probability by the understanding of the principles and by the success of computed tomography (CT),[1] different groups of investigators successfully applied the detection of NMR signals and, for PET, of annihilation radiation, to regional imaging in vivo, and, noninvasively, of nuclides (a) providing NMR signals, and (b) decaying with the emission of positrons. At the time I'am writing this paper (1985), both NMR imaging and PET have reached a considerable degree of maturity, in the sense that the physical properties governing these two technologies are fully understood and that the latter are applied with well-designed and reliable apparatus. The promises and the limitations of these approaches in biomedical research are now well understood.

PET was developed as a modality to yield regional information about biochemical processes essential to life. In this sense, it can be regarded as a natural extrapolation of the earlier investigation, usually by means of scintillation detectors, of some biochemical processes by the systemic administration of selected compounds labelled with "physiological" positron-emitting radionuclides. At no time was PET regarded as an imaging tool suitable for morphological studies. The history of NMR imaging differs somewhat from the evolution of PET.

---

[1] It should be noted that some of the early attempts in NMR and PET imaging do antedate that period

The enthusiasm of the chemists and biochemists utilizing NMR spectroscopy in the period between the late 1940s and the early 1970s was based, very justifiably, on the great usefulness of NMR as a probe for chemistry and in vivo biochemistry; the latter, among other applications, in the assessment of the chemical shifts of phosphorus compounds, which are of importance as a source of energy for metabolic processes. NMR imaging emerged in the early 1970s, however, as a morphological tool, mostly based on the imaging of proton distribution. Even at the present time, NMR imaging has made its main impact as a means of producing high-quality morphological images of the human body and of pathological conditions, based primarily on the detection of protons. Yet the clearly documented usefulness of NMR spectroscopy in an overwhelming number of chemical studies, many of them carried out in living systems, tantalizingly suggests that NMR imaging could incorporate the added dimension of spectroscopy with nuclides, other than protons, and thus provide a promising tool for the in vivo and regional determination of regional metabolism. Thus, on the basis of the above-mentioned premises, it is certainly reasonable to presume that PET and NMR imaging could in fact become competitive modalities for the achievement of the goals stated above.

In the assessment of the potential competition between PET and NMR imaging, two factors play dominant roles: the image-forming variables in PET and NMR imaging, and the resolutions achievable by the two methods.

In PET, the image-forming variable is always a radionuclide decaying by the emission of positrons and systemically administered to the subject under study in the form of a selected compound. Without the administration of the radionuclide, no PET image of a biological system can be obtained, because normally a biological system does not contain positron-emitting radionuclides in amounts sufficient for imaging.

For NMR imaging the situation is more complex, because this modality may and does utilize a number of the following imaging variables:

1. Distribution of nuclides which provide NMR signals
2. Distribution of relaxation times (spin-lattice and spin-spin) of "NMR nuclides"
3. Frequency shifts of NMR signals

4. Distribution of "NMR nuclide" tracers, incorporated into selected compounds
5. Distribution of NMR contrast media
6. Flow of fluids in and out of the field of view

NMR images can be obtained without the systemic administration of any nuclide, because biological systems do contain inherently a number of nuclides producing NMR signals, such as protons, ^{31}P, and ^{23}Na. Also, a number of others, or compounds containing such nuclides, can be administered systemically for NMR imaging. Furthermore, substances affecting NMR signals from naturally occuring nuclides (such as protons) can also be used as image-forming variables in NMR imaging.

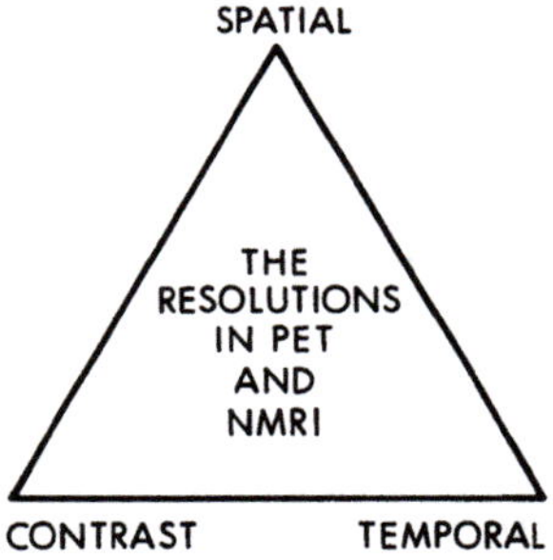

Fig. 1. Resulutions obtainable by PET and NMR imaging

In addition to the image-forming variables, the comparison of PET and NMR imaging, and, for that matter, the assessment of any imaging system, must also include an assessment of the resolutions (Fig. 1) which can be obtained for different imaging-forming variables. While it is most common to consider only the spatial resolution of an imaging system, it is also probably obvious that the contrast resolution (the ability of the system to distinguish small differences in spatial contrast) or temporal resolution (the ability of the system to resolve small differences in the image-forming variable in time) may also be of crucial importance in the assessment of the modality for a given purpose. The resolutions of any image-forming system depend on (a) the physical design of the device, and (b) the signal-to-noise ratio which can be achieved for a given image-forming variable. Within limits, any of the above resolutions can be optimized at the expense of another one.

In general terms, the signal-to-noise (S/N) ratio in most PET examinations is very high, because the only image-forming variable is provided by the administered radionuclide and the S/N ratio is, in general, limited predominantly by restrictions imposed upon the amount of radioactivity which can be administered. Similarly, in NMR imaging the S/N ratio is generally relatively poor, because the image-forming signal in NMR imaging is marred by considerable noise contributions from signals originating from spurious nuclides, thermal effects, magnetic field inhomogeneities, etc. This relatively poor S/N ratio in NMR imaging severely limits the resolutions of that modality, for nuclides which occur naturally or which are introduced into the field of view of the image.

An important difference between PET and NMR imaging is the use of tracer methodology. PET imaging relies exclusively on the use of radioactive tracers, which means that the distribution of a systemically administered radioactive indicator is followed, usually as a function of space and time. Then through the application of a suitable model, under favorable circumstances, a specific metabolic pathway can be unravelled from the information thus acquired. It is important to emphasize that in most instances tracer methodology demands that the chemical amount of the tracer material be small enough not to perturb the phenomenon under study. In most instances (see Table 1) the amount of radioactive

Table 1. Concentrations of Some Brain Materials and Specific Activities Required of Imaging Ligands (from McGeer [1])

	Tissue concentration mol/g	Specific activity mCi/mmol
Energy intermediates	10^{-6}–10^{-5}	10^{-1}–1
Amine neurotransmitters	10^{-9}–10^{-8}	10^{2}–10^{3}
Receptors	10^{-13}–10^{-11}	10^{5}–10^{7}

tracer administered, particularly in the case of the short-lived "physiological" positron emitters, easily fulfills the latter condition. However, in the study of some easily chemically saturated systems, for example neuroreceptors, even the quantities of material administered in PET imaging must be of particularly high specific activity (up to 1 Ci/μmol) and one must be prepared painstakingly to avoid chemical interference with the system under study. In the overwhelming majority of cases, tracer methodology per se cannot be applied in NMR imaging. This is because the relatively low S/N ratio yielded by this modality requires large chemical quantities of the NMR-signal-producing nuclides, and thus precludes their use as true tracers. Exceptions to this statement include situations where large amounts of a tracer material can indeed be administered without modifying the physiological process under study. For example, NMR angiography can be carried out by the replacement of a percentage of the circulating blood with a compound labelled with an NMR nuclide, such as a fluorocarbon labelled with ^{19}F or with an NMR signal inhibitor, such as gadolinium. This contrast manipulation of the NMR image can be likened to the use of radiographic contrast media (iodine, barium, air), and there is strong evidence at the present time that a number of such contrast materials will be developed for NMR and utilized effectively for the visualization of a number of organs. Nevertheless, this approach is not usually classified as tracer methodology, and any studies with subchemical quantities of tracer material appear to be beyond the scope of NMR.

The ability of NMR imaging to image certain nuclides which are the inherent components of living matter, a capability unchallenged by PET, is the fundamental basis upon which the usefulness and success of NMR imaging rests. Table 2 lists the principal natural components of living matter which exhibit NMR signals. This list also includes information on the natural abundance of these nuclides and a value for the

Table 2. NMR Imaging Index (from McGeer [1])

Isotope	Sensitivity ×	Abundance ×	Concentration Atoms/g × 10^{20} =	Imaging Index
^{1}H	1	100	460	46000
^{31}P	0.066	100	0.60	3.96
^{23}Na	0.09	100	0.42	3.78
^{14}N	0.001	100	6.00	0.60
^{13}C	0.016	1	24.0	0.38
^{35}Cl and ^{37}Cll	0.004	100	0.30	0.12
^{2}H	0.01	100	4.00[a]	
^{19}F	0.83	100	0.05[a]	
^{7}Li	0.29	100	0.14[a]	

[a] Not normally present in tissue; concentration required to achieve an imaging index of 4

strength of the signal which they may provide. There are several ways of expressing the relative NMR imaging sensitivity of various nuclides (such as sensitivity at constant field or sensitivity at constant frequency). An in-depth analysis of this subject is beyond the scope of this study, and Table 2 should be regarded only as a general indication of the order of NMR imaging sensitivities. It is quite evident that protons, which constitute a very large percentage of living matter, provide the most attractive image-forming variable for NMR imaging, at least from the S/N ratio point of view. At this time, for all practical purposes, NMR imaging relies upon the imaging of protons and of such associated variables as relaxation times and frequency shifts. Because of the strong S/N ratio achievable in the human body with NMR imaging of protons, the frequency shifts of the signals resulting from the chemical bonds of protons in various compounds (particularly water and lipids) provide a very promising method for distinguishing different compounds. It should be emphasized that this approach is impossible by PET because of the absence of positron-emitting isotopes of hydrogen; furthermore, PET is blind to chemical bonds. NMR imaging has also successfully produced images of the distribution of ^{31}P, ^{23}Na, and ^{19}F in the human brain. The sensitivity of the imaging of these nuclides is, however, considerably lower than that of protons, requiring serious compromises in spatial, temporal, or contrast resolution. For example, for protons at about 0.35 T state of the art NMR imaging devices yield a spatial resolution of approximately 14.4 (1.2 × 1.2 × 10 mm) mm^3/voxel in a period of a few minutes. If phosphorus imaging were to be carried out in the same period of time for the same S/N ratio,[2] the dimensions of the voxel would have to be between 1000 and 10000 times greater, depending upon the concentration of ^{31}P. This would lead to voxel sizes of between 14 and 140 cubic *centimeters,* corresponding to linear resolutions from 2.4–5.2 cm, respectively. This relatively pessimistic assessment should be moderated by the following considerations: it is reasonable to expect that the S/N ratio achieved in present systems will be improved in the foreseeable future by engineering progress, such as the use of higher and more uniform magnetic fields and the optimization

---

[2] In this comparison, equal S/N ratios may not be the optimal criterion. For example, in many instances of imaging, S/N ratio might be profitably traded for a gain in contrast. Nevertheless, the above comparison is adequately valid as a general example

of receiver coil designs. It is impossible to predict with any certainty what gains can be thus achieved, but it is doubtful that they will yield improvement of the S/N ratio greater than a factor of 20, leading to an improvement in the linear spatial resolution, under very favorable circumstances, by a factor of about 2.7. Another important comment on this example is that even the relatively poor spatial resolution of NMR imaging of phosphorus and sodium has already been shown to yield quite useful physiological information in the imaging of the distribution of these nuclides in the adult brain and in the excellent studies of the metabolism (with ^{31}P) of the brain studied in the whole infant organ. Also, improvements in the S/N ratio can be achieved through the use of surface coils, but with a limited depth perception.

In contrast to the above, PET measurements of cerebral metabolism (oxygen and glucose) and of myocardial metabolism (palmitic acid and glucose) are now carried out with a spatial resolution of between 500 and 1000 mm^3/voxel in a period of time which varies between 40 s and a few minutes, and with an S/N ratio which allows a contrast resolution of better than 10%. There is evidence substantiated by experimental results that this spatial resolution will be reduced to less than 250 mm^3 within the next 2 years. On the other hand, the absence of suitable radionuclides does not permit the determination by means of PET of the distribution of either phosphorus or sodium (^{22}Na, which decays with the emission of positrons with a half-life of about 2.6 years, is too long-lived, and ^{30}P, also a positron emitter, decaying with a half-life of 2.5 min, is probably too short-lived for the effective measurement of phosphorus metabolism by PET).

From the above considerations it is my opinion that NMR imaging and PET occupy within the biomedical imaging armamentarium places which are distinctly and widely separated by the type of information they can provide. Proton imaging by NMR yields morphological information unchallenged by PET. Furthermore, the utilization of other variables in proton imaging, such as relaxation times and frequency shifts, adds potentially highly useful physiological information. The development of NMR contrast materials opens up very promising avenues in applications generally similar to that of contrast materials in conventional radiology. The imaging of other nuclides with NMR signals, particularly such as ^{31}P and ^{23}Na, already shows some clear indications of clinical usefulness. NMR spectroscopy of compounds containing ^{31}P has already proven to be highly useful in the assessment of phosphate-mediated metabolism, particularly in smaller accessible structures with a favorable S/N ratio situation. There are strong indications that phosphorus frequency spectroscopy will be applicable, albeit with a relatively poor spatial resolution, in larger organs. The unfavorable S/N ratio inherent in NMR imaging presents a formidable barrier to the improvement of resolutions of this modality other than for proton imaging. At this time, the understanding of NMR technology does not predict improvements in the S/N ratio by increments greater than about one order of magnitude.

PET imaging occupies an unchallenged position for the in vivo and regional assessment of metabolism and associated parameters by tracer methodology. The number and application of compounds of potential use in biomedical research labelled with positron-emitting radionuclides now exceeds several hundred, and is rapidly growing. The S/N ratio in PET is inherently high, and the understanding of this technology allows the reliable prediction that the resolutions in this modality (spatial, contrast, temporal) will increase several-fold in the safely predictable future. Nevertheless, these resolutions are limited by physical factors and by the amount of radioactivity that can be safely administered for diagnostic purposes.

Our present understanding of NMR imaging allows the prediction that in the foreseeable future, NMR imaging will have a profound effect on the delivery of health care through the broad utilization of this modality in clinical centers. It is quite probable that the dissemination of the number of PET centers will be limited to large medical centers, because of the complexity and cost of this methodology. However, it is not irrational to speculate that the ultimate effect of PET on medicine may be farther reaching than that of NMR imaging, because PET allows the probing of some of the fundamental factors governing physiology and pathophysiology.

Reference

McGeer PL (1983) The Role of Imaging in Health Care — a hook at the Future. In: Proceedings of the National Conference on Biological Imaging, II Clinical Aspect, National Academy of Sciences, Washington DC, Oct 1983, p 186

Aims on Phosphorus-31 Magnetic Resonance Imaging

K. Kogure, H. Ohtomo, S. Matsui, and H. Kohno

Introduction

Nuclear magnetic resonance (NMR) is capable of producing anatomical and biochemical data on the brain. Hydrogen-1 [11, 15] exists in abundance in the human body, differs in concentration from one tissue to another, and can be readily detected. Many investigators have therefore used it in tomography [8, 9, 12, 21], and the technique has reached a stage of practical application. However, since only limited biochemical data are available for ^{1}H, the application of ^{13}C, ^{19}F, ^{31}P, and others is being tried. Using ^{31}P-NMR, changes in phosphocreatine, adenosine triphosphate (ATP), sugar phosphates and inorganic phosphate can be determined, but it takes 30–60 min to produce an image. Therefore, ^{31}P-NMR is inadequate for tomography at present. Thus, our aim is to obtain the local spectra of ^{31}P-NMR [1, 13, 26] in vivo in a specified region, utilizing ^{1}H-NMR-CT (computed tomography) as guide, and to establish a biochemical diagnosis of ischemic lesions in the brain. For this purpose, we conducted a fundamental study, in which alterations in phosphorus metabolism were investigated by ^{31}P-NMR spectrometry of the rat ischemic brain and the hypoxic brain, as shown in the sections "In Vitro Study" and "In Vivo Study." In addition, we are now working on ^{31}P-MRI as shown in the section headed "MRI of ^{31}P". These works will be mentioned in this order.

In Vitro Study

Methods

Male Wistar rats weighing 250–300 g were used. As shown in Fig. 1, both vertebral arteries were coagulated to block the blood flow by Pulsinelli and Brierley's [25] method after anesthetization by an

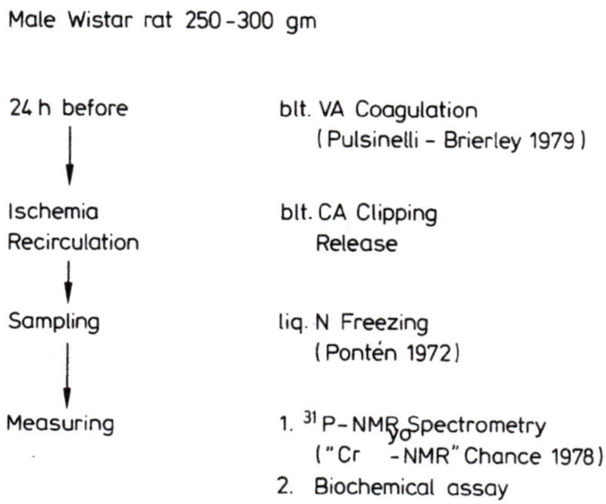

Fig. 1. Schematic of the procedure in the experiment. *VA*, vertebral artery; *CA*, carotid artery; *BLT*, bilateral

intraperitoneal injection of pentobarbital. After 24 h, the animals were immobilized with tubocurarine following induction of ether anesthesia. Next, the blood flow in both common carotid arteries was blocked and reopened with micro-Scoville clips, following heparinization under

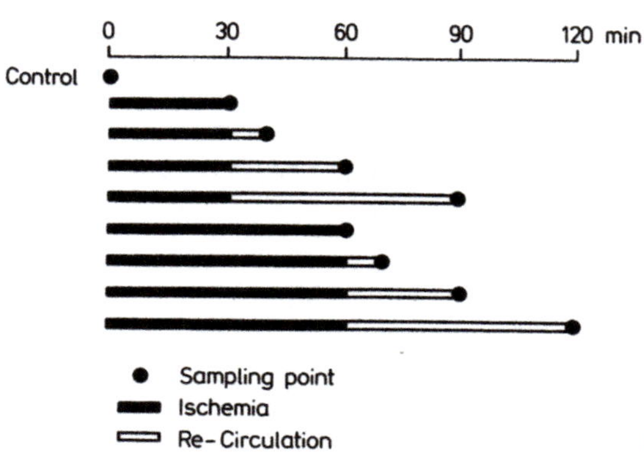

Fig. 2. Program for sampling

controlled respiration (Harvard Rodent Respiration Apparatus 680), according to the program shown in Fig. 2. Changes in phosphorus metabolism were investigated 0, 10, 30, and 60 min after reperfusion, following 30- and 60-min ischemia. The experiment was carried out in a stable state, i.e., a rectal temperature of 37.0 °C, respiratory rate of 56/min, PaO_2 at 35–45 torr, pH of 7.30–7.45, and mean blood pressure of not lower than 100 mmHg. Sampling was done by Pontén's method [24] as follows: we froze the brain by pouring liquid nitrogen over the skull, from which the scalp epicranium had been peeled off, and removed it in a frozen state after 10 min, when the distortion caused by the freezing technique had disappeared. The freeze-trapped rat brain [6], which was ground in a mortar filled with liquid nitrogen, was put into 10-mm NMR tubes, and ^{31}P-NMR spectra were accumulated with a Bruker CXP-300 FT-NMR at 121.5 MHz and −5 °C in a nonspinning sample tube. The temperature was automatically stabilized. Then proton irradiation

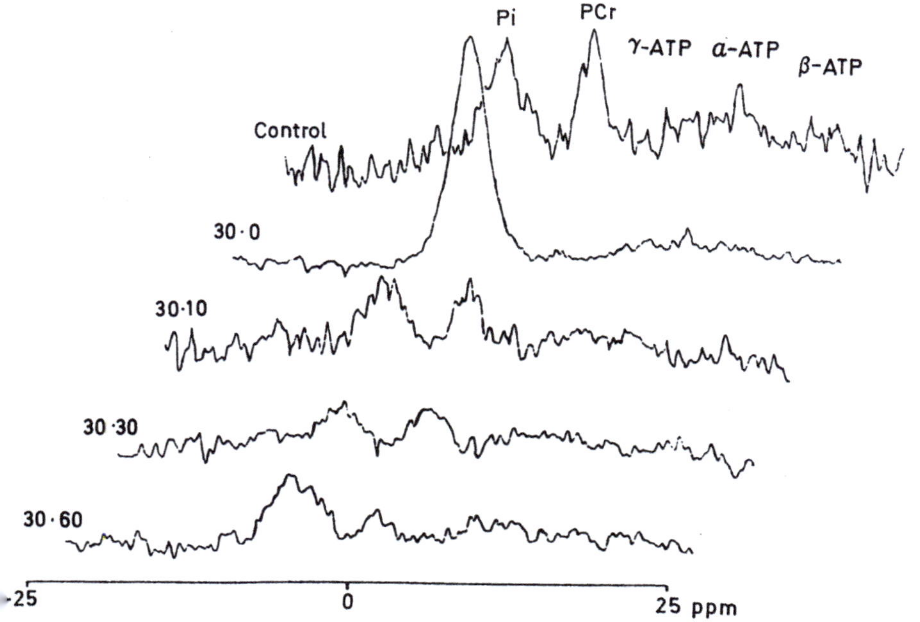

Fig. 3. ^{31}P-NMR spectrograms obtained 0, 10, 30, and 60 min after reperfusion following 30 min of ischemia

(2 W) was applied to increase the signal-to-noise (S/N) ratio. An equipped transmitter provided 90° pulse widths of 30 μs for the nucleus at a peak-to-peak voltage of 300 V. (Typical spectra were measured by using a 30° pulse). [Note: The signal of ^{31}P cannot be taken below −20°C, where ATP still is decomposed by adenosine triphosphatase (ATPase) for a short time, but the peak of inorganic phosphorus (Pi) cannot be demonstrated.]

Results

1. As shown in Fig. 3, resynthesis of ATP occurred and phosphocreatine (PCr), used as a substrate in ATP synthesis, was detected in large quantities after reperfusion following 30 min of ischemia. In contrast, after 60 min, resynthesis of ATP did not occur, the PCr peak was barely noticeable, and only Pi was present in large amounts, which

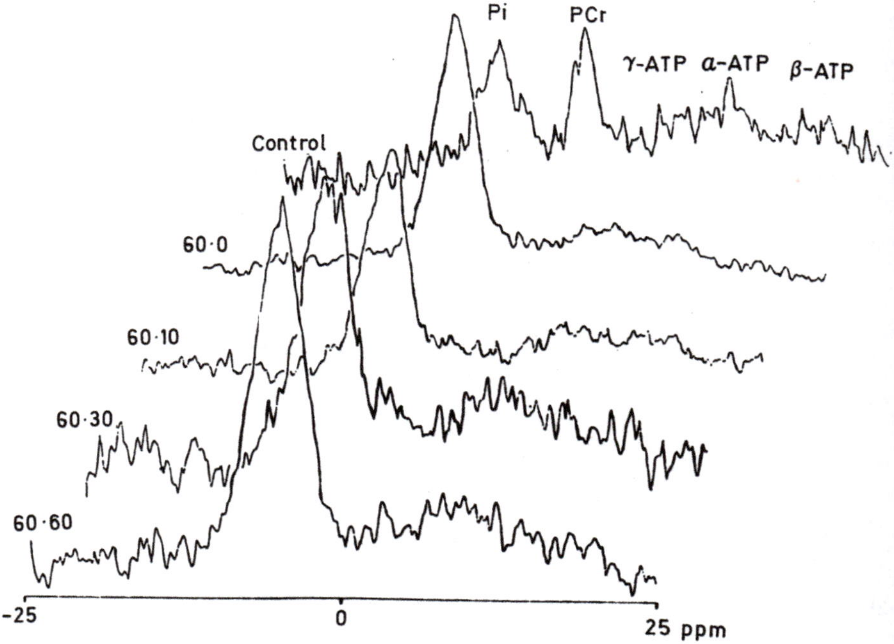

Fig. 4. ^{31}P-NMR spectrograms obtained 0, 10, 30, and 60 min after reperfusion following 60 min of ischemia

were even higher than the control values (Fig. 4). The tendency for PCr and ATP to be resynthesized after reperfusion of 30 min was also proved by ordinary biochemical determinations (fluorometric method) in rat brains subjected to 30-min ischemia alone, or followed by 10-min reperfusion, as shown in Table 1. PCr rose from 0.00 to 2.20 m*M*/kg, and ATP from 0.04 to 1.30 m*M*/kg.

Table 1. PCr, ATP, adenosine diphosphate (ADP), and adenosine monophosphate (AMP) levels determined by the fluorometric method in the control group and in the group subjected to 30-min ischemia alone or followed by reperfusion for 10 min[a]

	Control	30-min Ischemia	10-min Re-circulation
PCr	5.30	.00	2.20
ATP	3.00	.04	1.30
ADP	.03	.09	.05
AMP	.003	.28	.06

[a]Amounts shown are in m*M*/kg

2. Because the chemical shift on the NMR-spectrogram is greatly influenced by pH [20], the shift of Pi from normal phosphoric acid was expressed as a function of pH (Fig. 5). As a result, it was demonstrated that pH had returned approximately to a normal level 10–30 min after reperfusion following 30 min of ischemia, but not in the group subjected to 60 min of ischemia. This demonstrates that acidic tissues begin to

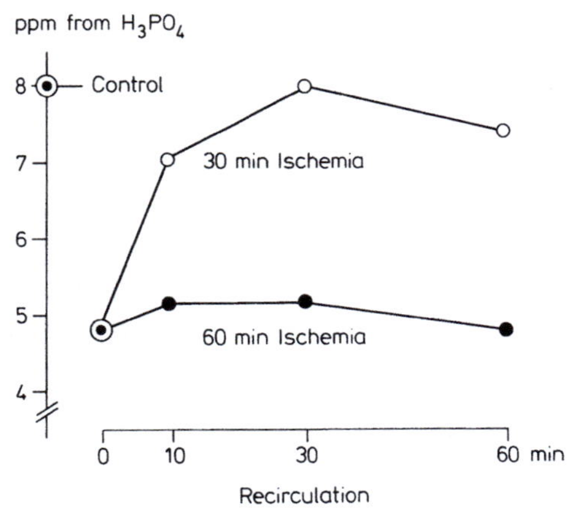

Fig. 5. Chemical shifts of inorganic phosphate from normal phosphoric acid (H_3PO_4), 0, 10, 30, and 60 min after reperfusion in the 30-min (*open circles*) and 60-min (*closed circles*) ischemia groups

respire again with reperfusion and can utilize excess acids in the brain which has been subjected to 30-min ischemia, and suggests that pH determination by NMR may be used as an indicator of the extent of recovery of metabolically injured tissues. The ratios of the peak heights of PCr to those of Pi on the spectrogram are plotted as an indicator of energy on the ordinate, with open circles for 30-min ischemia and closed circles for 60-min ischemia in Fig. 6. The ratio recovered to the normal level 10–30 min after reperfusion, and was slightly lower after 60 min.

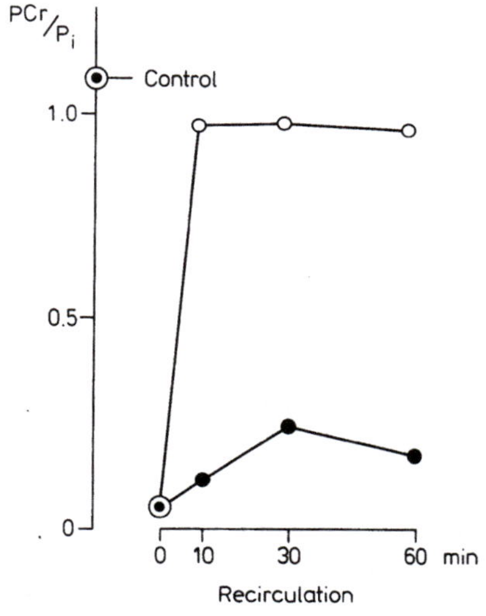

Fig. 6. Ratios (PCr/Pi) of peak heights of phosphocreatine (PCr) to those of inorganic phosphate (Pi), 0, 10, 30, and 60 min after reperfusion in the 30-min (*open circles*) and 60-min (*closed circles*) ischemia groups

Discussion

The results suggest that disturbances of tissue respiration in cerebral ischemic lesions may exceed the critical point for the damage which lies between 30- and 60-min ischemia in this model. Reperfusion following 30, 40 (Fig. 7), and 60 min of ischemia showed progressive decrements, mainly in the ATP and PCr peaks, indicating profound, presumably irreversible, tissue damage at this time.

In 1978, Chance et al. [5, 6] carried out ^{31}P-cryo-NMR determination at 72.9 MHz and −10°C for the whole brain after rapid freezing with liquid nitrogen under pentobarbital anesthesia.

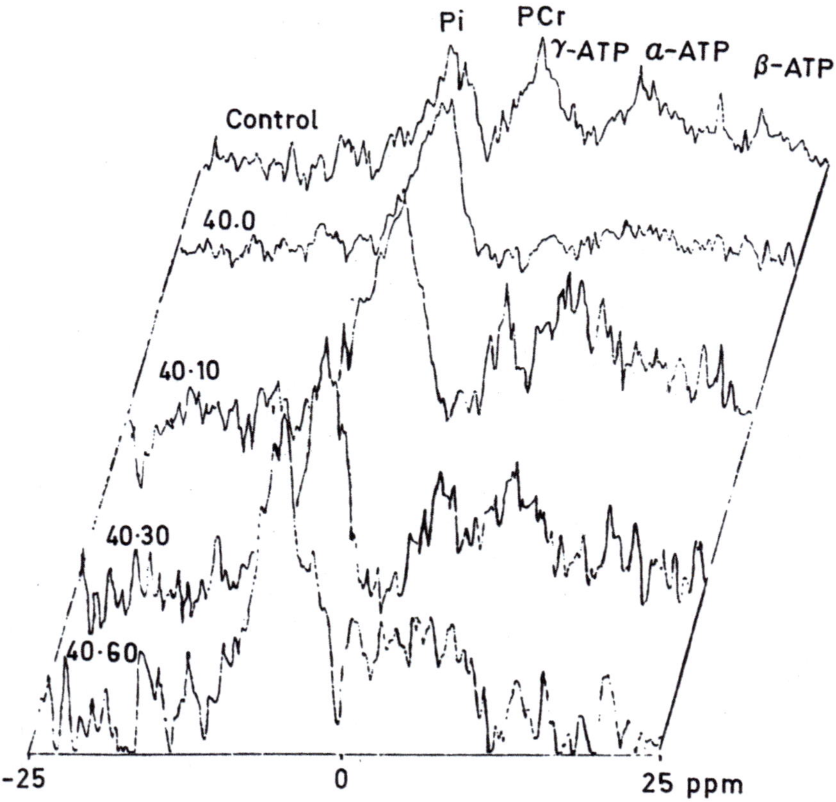

Fig. 7. ^{31}P-NMR spectrograms obtained 0, 10, 30, and 60 min after reperfusion following 40 min of ischemia

We have investigated changes in phosphate compounds using the same method in models of whole cerebral ischemia of varying duration. The investigation will be meaningful as a basis for in vivo studies in ^{31}P-topical-NMR and its clinical application [9].

In the present study, we are confronted with various unsolved problems, such as establishment of the temperature at which determination should be done; alterations with time in energy metabolism, which are caused by changes in temperature; and difficulty in the measurement of solid samples with NMR apparatus, etc. We are planning to conduct further studies with respect to these points.

In Vivo Study [Tomographic Magnetic Resonance (TMR)]

Hydrogen-1-NMR imaging allows dynamic studies of the rat brain, since H_2O shifts rapidly follow physiological changes. In pilot studies on cortical cryoinjuries and/or hypoxemia, images showed the same location and extent of cell damage as did histology. It appears this technique can define areas of damage following stroke sooner than computerized axial tomography (CAT) scans.

Using an NMR surface coil technique we were able to obtain high spectral resolution of ^{31}P compounds in the rat brain (typically 30 min per measurement). The measurements gave relative concentrations of various phosphate metabolites: PCr, ATP, sugar phosphate, and Pi.

Phosphorus-31-NMR spectra were recorded at 72.88 MHz using a Bruker WH-180 spectrometer (Bruker Instruments) with a 90-mm-bore superconducting magnet. The surface coil was an 11-mm diameter, two-turn flat spiral, turned on an especially constructed circuit board and probe insert, and mounted on a modified Bruker NMR probe body [23].

The biochemical assays for PCr, ATP, glucose (Glu), glucose-6-phosphate (G-6-P), pyruvate (Py), and lactate (La) were performed on groups of similarly prepared animals by direct fluorometric methods [17]. Focal brain tissue for those assays was sampled from the area corresponding to the effective field of the surface coil by Pontén's [24] freeze in situ technique, and extracted by the method of Brunner et al. [4].

The S/N ratio for these data was improved by incorporating proton decoupling [2]. Quantitation is complicated by a broad baseline hump, which is derived from bone phosphorus [1] and an instrumental artifact. Various tissue depths can be probed by varying the NMR pulse duration, and this technique was applied over necrotic tissue produced by cryoinjury. Data obtained by this and traditional biochemical methods show agreement for normoxia, severe hypoxia, partial necrosis, and partial ischemia. This technique provides a means of monitoring the treatment of stroke, recovery of an injured site, and/or physiological chemistry in vivo.

Figure 8A is a chronological series of localized spectra obtained from one rat. Spectra *a* and *b* were obtained in the periods 5–38 min and 60–93 min after insertion of a normal animal into the magnet. The peak assignment and chemical shifts in parts per million (ppm) are denoted.

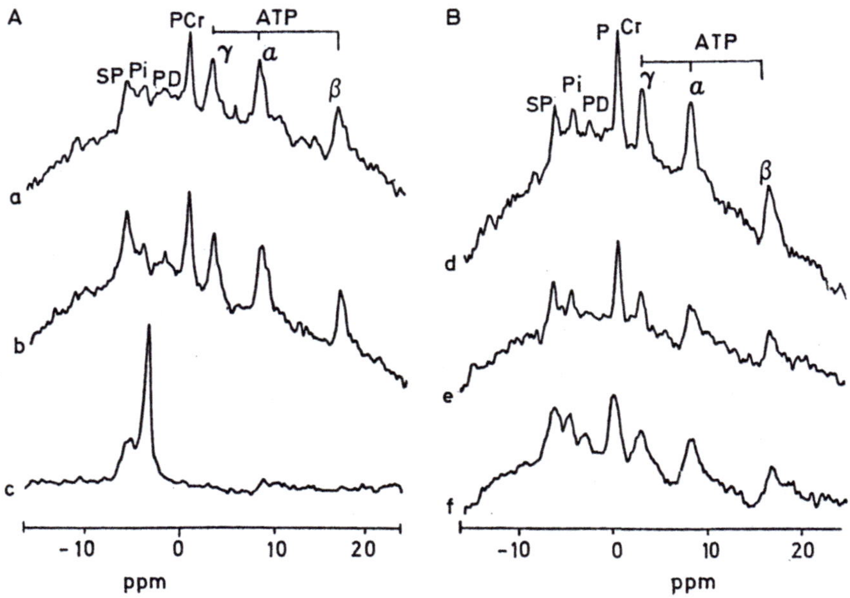

Fig. 8 A, B. ^{31}P-NMR spectra. In **A,** *a* and *b* are spectra of a normal rat brain at 5–38 min and 60–93 min after the rat was inserted in the magnet; *c* shows the effect of severe hypoxemia. *PD,* phosphodiesters; *SP,* sugar phosphates. **B**, spectra obtained from a partially necrotic (*d, e*), and a partially ischemic (*f*) rat brain. (*d* and *f*, 20 μs and *e*, 6 μs pulse length)

Table 2. Biochemically assessed levels of cerebral energy metabolites in rats during steady-state control condition. Values are means + standard error of mean (SEM) [m*M*/kg]

	PCr	ATP	ADP	AMP	Glu	G-6-P	Py	La
Control	5.34	2.83	.345	.012	2.67	0.12	.097	1.24
n = 6	±.08	±.05	±.021	±.007	±.49	±.009	±.012	±.06
Hypothermia[a]	5.33	2.88	.331	.020	4.03*	0.20*	.086	0.87*
n = 5	±.12	±.04	±.017	±.008	±.38	±.008	±.010	±.05
Hypoxemia[b]	0.15*	0.07*	.092*	.083*	0.49*	0.07*	.016*	23.2
n = 5	±.006	±.002	±.008	±.040	±.03	±.007	±.008	±2.1

n, number of observations; *, significantly different from control values [22]

[a]After 20 min of hypothermia (body temperature 30 ± 2°C)
[b]After 20 min of hypoxemia (PaO_2 = 13 ± 3 torr)

Contributions to the α- and γ-phosphate peaks of ATP from adenosine diphosphate (ADP) in Fig. 8 spectra were undetectable. Spectrum *c* shows dramatically the effect of severe hypoxemia [13]. Table 2 shows complete conversion of PCr and ATP to Pi.

Figure 8B depicts the effects of focal cryonecrosis in the cortex (Table 3), and of altering the NMR pulse length to vary the depth of sensitive volume. In *d*, the pulse length was 20 μs, indicating a sensitive volume located in a region at a depth of about 3–8 mm. The spectrum is consistent with the assertion that little necrotic tissue resides in the sensitive volume. In *e*, the pulse length was 6 μs, reducing the sensitive volume depth to approximately 0–5 mm. Spectrum *f* (pulse length, 20 μs) represents the effect of carotid artery ligation (partially ischemic rat brain) [15]: the Pi level is significantly elevated relative to the normal spectra, and contributions from normoxic regions result in relatively high levels of the other metabolites.

Table 3. Biochemically assessed levels of cerebral energy metabolites 30 min after cryoinjury, and 20 min after the beginning of levine's preparation [16]. Values are means + SEM (m*M*/kg)

	PCr	ATP	ADP	AMP	Glu	G-6-P	Py	La
Cryoinjury (n = 5) injured hemisphere	4.72 ±.16	2.27 ±0.05	.278 ±.022	.022 ±.008	4.02* ±0.60	0.18 ±.010	.116 ±.015	2.10* ±0.16
Intact hemisphere	5.30 ±0.20	2.76 ±0.07	.334 ±.026	.014 ±.008	3.90 ±0.49	0.15 ±.009	.110 ±.010	±0.74 ±0.17
Levine's rat (n = 5) ligated hemisphere	3.65* ±0.18	1.80* ±0.03	.357 ±.019	.103 ±.011	1.25* ±0.27	0.06* ±.010	.032* ±.005	18.4 ±1.22
Patent hemisphere	4.00 ±0.10	2.70 ±0.05	.300 ±.010	.072 ±.006	5.35* ±0.72	0.18 ±.010	.079 ±.008	5.61* ±0.77

n, number of observations; *, significantly different from the control values (see Table 2)

Hypoxemia, cryoinjury, and ischemia produced a decrease in high-energy metabolite levels and generally elevated sugar phosphate and lactate levels according to the NMR results (Fig. 9).

One can follow brain trauma (by edema, hemorrhage, etc.) with rapid ^{1}H-NMR imaging, and then apply high-resolution ^{31}P-NMR

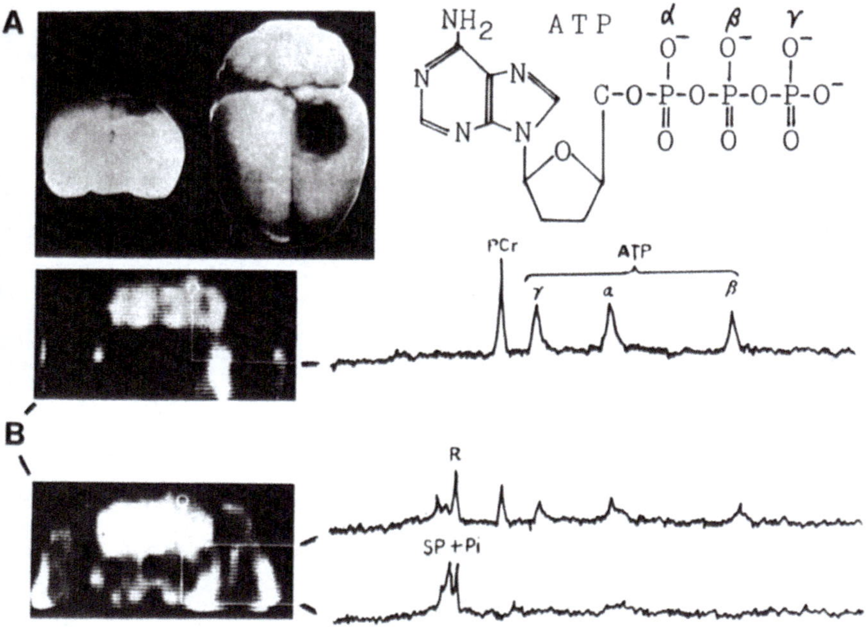

Fig. 9 A, B. **A** Cryoinjured brain and its coronary section. **B** ^{1}H-NMR image and ^{31}P-NMR spectrogram on a brain region, *Upper,* normal brain; *lower*, perifocus (*upper*) and necrotic focus (*lower*) of the cryoinjured brain

spectrography for regional phosphate metabolism, as shown in Fig. 10. One instrument should provide both types of data, though none have been reported. This is due to different optimal magnetic field requirements for ^{1}H and ^{31}P. High-resolution NMR coil work uses liquid-helium-cooled super-conducting coils; it is difficult to change their field strength. Phosphorus-31 field strengths are too high for imaging, and a solution to this problem must be found before a dual instrument can be built.

Magnetic Resonance Imaging (MRI) of ^{31}P

Some studies on magnetic resonance imaging (MRI) of ^{31}P have already been reported [19]. The following points remain as problems [4, 18]: long data acquisition times (about 1 h), the low S/N ratio, the necessity

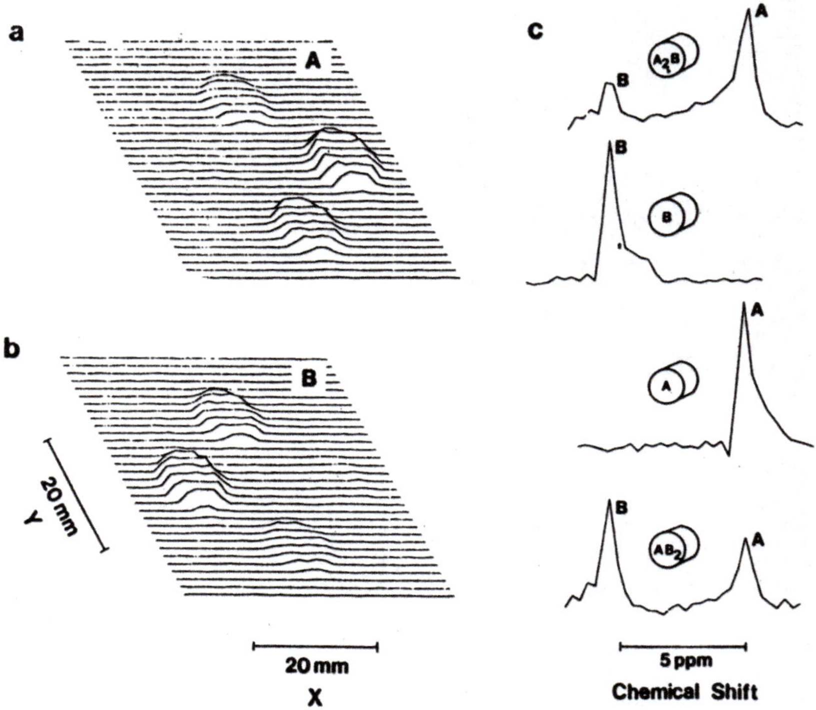

Fig. 10 a–c. Proton images of acetone **a** and benzene **b**, and proton high-resolution spectra at four different locations (**c**). Both the X and Y axes are represented by 32 points, and the spectra by 32 points, interpolated from 16 points using the zero-filling technique. The images were obtained from the spectral intensities of absolute values of the Fourier transforms, while the local spectra are of phase-corrected absorption mode. [18]

of higher magnetic field strengths and wider spectral bandwidths, the reproducibility of the spectroscopy. The solutions for these problems will lead to the establishment of chemical shift imaging techniques.

More advanced techniques of chemical shift imaging can produce strict image of each energy metabolite (PCr, Pi, ATP). Accordingly, we performed experiments on a high-speed method of spatially resolved high-resolution NMR spectroscopy for biological systems and obtained the results set out below [1].

Experiments were performed at 0.5 Tesla operating our ^{1}H-NMR imager. Spin echo trains induced by periodical inversions of the field gradient were further phase-modulated by pulse applications of other field gradients.

Such phase-modulated echo trains can be converted to three- or four-dimensional data, including both spectroscopic and spatial information, by means of suitable data manipulation involving echo rearrangement and multidimensional Fourier transformation. Experimental demonstration has been made for proton-NMR, using a two-dimensional test sample consisting of solutions of acetone and benzene (Fig. 10); the ^{1}H NMR chemical shift difference is 5.1 ppm. The spectral resolution of each local spectrum is limited solely by the corresponding local field inhomogeneity. Molecular motion, however, may disturb the echo formation, leading to spectral broadening in a manner similar to the rotational echo train [18]. Nevertheless, this broadening can be used to detect one-dimensional slow diffusions, specifically along the direction of the inverted field gradient. The reciprocal of the inversion period also defines the spectral bandwidth (200 Hz in this case). For applications to other nuclei, such as ^{31}P, ^{13}C, ^{15}N, etc., a greater bandwidth will be needed at a higher magnetic field. Consequently, a new technique is also suggested for inherently increasing the spectral bandwidth, which is limited by the inversion period of the field gradient. The application of this technique permitted a spectral bandwidth of 40 ppm or even a freely expanded one [17, 18]. Utilizing a chemical shift imaging technique in this band, we are now developing an NMR imager for ^{31}P. This technique may provide information about the tissue pH in vivo as well as the tissue energy state. MRI of ^{31}P is currently less practical, but a sparkling future for MRI technology of ^{31}P seems assured.

References

1. Ackerman JJH, Grove TH, Wong GG, Gadian DG, Radda GK (1980) Mapping of metabolites in whole animals by ^{31}P-NMR using surface coils. Nature 283:167–170
2. Bottomley PA, Nunnally RL (1980) Localized ^{31}P-NMR studies of the metabolism of perfused hearts: new methods for assessing drug therapy. Bull Magn Reson 2:423–424
3. Bottomley PA, Foster TH, Lene WH (1984) In vivo nuclear magnetic resonance chemical shift imaging by selective irradiation. Proc Natl Acad Sci USA 81d:6856–6860
4. Brunner EA, Passonneau TV, Molstad C (1971) The effect of volatile anaesthetics of levels of metabolites and on metabolic rate in brain. J Neurochem 18:2301–2316
5. Chance B, Eleff S, Leigh JS, Sokolow D (1980) Noninvasive, nondestructive approach to cell bioenergetics. Proc Natl Acad Sci USA 77:7430–7434
6. Chance B, Nakase Y, Bond M, Leigh JS, McDonald G (1978) Detection of ^{31}P nuclear magnetic resonance signals in brain by in vivo and freeze-trapped assays. Proc Natl Acad Sci USA 75:4925–4929

7. Doyle FH, Gore JC, Pennock JM, Bydder GM, Orr JS, Steiner RE, Young IR, Burl M, Clow H, Gilderdale DJ, Bailes DR, Walters PE (1981) Imaging of the brain by nuclear magnetic resonance. Lancet II:8237–8241
8. Edelstein WA, Hutchison JMS, Smith FW, Mallard J, Johnson G, Redpath TW (1981) Human whole-body NMR tomographic imaging: normal sections. Br J Radiol 54:149–151
9. Edwards RHT, Dawson MJ, Wilkie DR, Gordon RE, Shaw D (1982) Clinical use of nuclear magnetic resonance in the investigation of myopathy. Lancet I:725–730
10. Hassen G, Crooks LE, Davis P, De Groot J, Herfkens R, Margulis AR, Gooding C, Kaufman L, Hoenninger J, Arakawa M, McRee R, Watts J (1980) In vivo imaging of the rat anatomy with nuclear magnetic resonance. Radiology 136:695–700
11. Herfkens R, Davis P, Crooks L, Kaufman L, Price D, Miller T, Margulis AR, Watts J, Hoenninger J, Arakawa M, McRee R (1981) NMR imaging of the abnormal live rat and correlation with tissue characteristics. Radiology 141:211–218
12. Hoult DI, Busby SJW, Gadian DG, Radda GK, Richards RE, Seeley PJ (1974) Observation of tissue metabolites using ^{31}P nuclear magnetic resonance. Nature 252:285–287
13. Kogure K, Scheinberg P, Utsunomiya Y, Kishikawa H, Busto R (1977) Sequential cerebral biochemical and physiological events in controlled hypoxemia. Ann Neurol 2:304–310
14. Lauterbur PC (1973) Image formation by induced local interactions: examples employing NMR. Nature 242:190–191
15. Levine S (1960) Anoxic-ischemic encephalopathy in rats. Am J Pathol 36:1–16
16. Lowry OH, Passonneau JV (1972) A flexible system of enzymatic analysis. Academic New York
17. Matsui S, Sekihara K, Kohno H (1985) High-speed spatially-resolved high-resolution NMR spectroscopy. J Am Chem Soc 107:2817–2818
18. Matsui S, Sekihara K, Kohno H (1985) High-speet spatially resolved NMR spectroscopy using phase-modulated spin-echo trains. Expansion of the spectral bandwidth by combined use of delayed spin-echo trains. J Magn Reson (in press)
19. Maudsley AA, Hilal SK, Simon HE, Witterkock S (1984) In vivo MR spectroscopic imaging with P-31. Radiology 153:745–750
20. Moon RB, Richards JH (1973) Determination of intercellular pH by ^{31}P magnetic resonance. J Biol Chem 248:7276–7278
21. Moore WS, Holland GN (1980) Nuclear magnetic resonance imaging. Br Med Bull 36:297–299
22. Nilsson L, Kogure K, Busto R (1975) Effects of hypothermia and hyperthermia on brain energy metabolism. Acta Anaesthesiol Scand 19:199–205
23. Nummally RL, Bottomley PA (1981) A phosphorus nuclear magnetic resonance assessment of pharmacological treatment of myocardial infarction using surface coils. Science 211:177–180
24. Pontén U, Ratcheson RA, Salford LG, Siesjö BK (1973) Optimal freezing conditions for cerebral metabolites in rats. J Neurochem 21:1127–1138
25. Pulsinelli WA, Brierley JB (1979) A new model of bilateral hemispheric ischemia in the unanesthetized rat. Stroke 10:267–272
26. Radda GK, Chan L, Bore PB, Gadian DG, Ross BD, Styles P, Taylor D (1982) Clinical applications of ^{31}P-NMR. In: Karstaedt N, Witcofski RL (eds) Proc International NMR Symposium, Bowman-Gray School of Medicine, October 1–3, 1981. Bowman-Gray School of Medicine

In Vivo Nuclear Magnetic Resonance Imaging of Sodium-23 in the Human Head*

S. K. Hilal, A. A. Maudsley, J. B. Ra, H. E. Simon,
P. Roschmann, S. Wittekoek, Z. H. Cho, and S. K. Mun

Introduction

In vivo images of tissue sodium were reported previously [7, 8, 12] using ^{23}Na observations of the normal cat's head and of cats with experimentally produced stroke. We present in this report the first images of sodium distribution in human subjects, which were announced recently [6].

Sodium is primarily distributed in the extracellular space. An increase in sodium concentration in tissues occurs when the extracellular space expands, as in edema, or when there is intracellular diffusion of sodium, as in the case of damage to the cell membrane. Besides edema and cell damage, sodium ion fluxes in animal cells play an important role in a variety of fundamental processes of cell activity, such as the control of mitogenic and oncogenic phenomena. Intracellular sodium concentration in malignant cells increases to 300% of the values measured in corresponding normal cells [3]. In rapidly dividing cells, sodium concentration reaches 150% of the value measured in slowly dividing cells [3]. Potentially, sodium imaging in tumors could provide an index of mitotic activity. It is important to emphasize that sodium distribution is not identical to proton distribution, and that they can each vary independently in a multitude of physiologic and pathologic conditions.

Biochemical Background of Sodium Imaging

An evaluation of NMR sodium imaging is facilitated by a consideration of the relative sensitivity and relative contrast of nuclear magnetic

* This paper was originally published in the Journal of Computer Assisted Tomography 9:1–7 (1985).

resonance (NMR) imaging for both protons and sodium in living tissues in normal and diseased states. In the brain the normal concentration of ^{23}Na is 0.14–0.16 *M* in the extracellular space [9, 11, 13] and 0.012–0.02 *M* in the intracellular space. The average concentration of sodium in a volume of brain tissue that comprises both intra- and extracellular compartments is approximately 0.045 *M*. The extracellular space in the brain represents 8%–20% of total volume [11, 13]. These concentrations of sodium in brain tissue are to be compared with a proton concentration of approximately 85 *M*. In brain tissues there are, therefore, 2000 times more protons than ^{23}Na atoms. The decreased sodium abundance and other factors cause the pronounced reduction of the sodium signal compared with that of protons. The variation of sodium concentration from tissue to tissue and the changes seen as a result of pathologic lesions are, however, more pronounced that those with protons. For example, in the case of brain edema caused by the expansion of the extracellular space only, without cell damage, an increase of tissue water by 10% will result in a 50% increase in sodium.

A quantitative study of the relative concentration of sodium and protons in brain tissue is discussed for the purpose of evaluating sensitivity and contrast. The concentration of protons and sodium in the normal brain and ventricles (cerebrospinal fluid, CSF) and in infarction was measured repeatedly by many previous investigators [9, 10, 11, 13]. These results are summarized in Table 1.

Image contrast (C) is commonly derived as follows:

$$C = (I_o - I_B)/I_B$$

where I_o is the signal intensity received from the object of interest and I_B is the signal intensity received from the background. The expected image contrast can be calculated from the concentration of protons and

Table 1. Concentration of protons and sodium in brain tissue

Tissue	Proton (*M*)	Sodium (*M*)
Normal brain	85	0.045
Ventricles (cerebrospinal fluid)	110	0.150
Infarction (maximum measured values)	96	0.150

sodium atoms listed in Table 1. The values of the anticipated image contrast are shown in Table 2.

Table 2. Values of anticipated image contrast

	Image contrast	
Tissues under comparison	Proton	Sodium
Ventricles/normal brain	0.29	2.3
Infarctions/normal brain	0.13	2.3

These results show that the brain/ventricle contrast is 7.9 times greater on sodium images than on proton images, and that the infarction/brain contrast is 18 times greater on sodium images than on proton images. In this discussion we are assuming that the contrast is a function of the spin density alone. Although this is substantially true in the case of sodium imaging, proton imaging relies also, in fact more markedly, on T_1 and T_2 contrast. These relaxation constants in biological tissue are not predictably correlated to proton concentration and are not, therefore, a measure of water abundance.

The greater image contrast expected with sodium is further accentuated above the values listed in Table 2 by the fact that intracellular sodium in the normal cell is less "NMR-visible" than in the CSF or in the extracellular spaces, because of the much shorter T_2 of intracellular sodium [1, 4]. The sodium signal arising from normal brain is, therefore, lower than that expected from the concentrations listed in Tables 1 and 2, thus leading to a higher image contrast. A possible explanation of the variation in the "NMR visibility" of sodium may be found, in part, in the T_2 relaxation characteristics of sodium. Although protons and other nuclei most frequently observed in biological systems have a nuclear spin of ½, relaxing with a single NMR transition, sodium has a spin of 3/2 and a quadrupole moment. Sodium can potentially have three transitions and two relaxation rates. Although in dilute aqueous solutions sodium has a single T_2 relaxation rate of approximately 50 ms, in living tissue it appears to have at least two transverse relaxation times, a long T_2 of approximately 16–30 ms and a short T_2 of 0.7–3.0 ms. In our laboratory we measured two T_2 values for sodium in the cat's head at 30 MHz, namely, 30 and 3 ms. In isolated living tissue, the short T_2 fraction represents 60% of all sodium [1, 4, 5]. The short T_2 relaxation rate of sodium decreases the observed signal from normal brain.

Materials and Methods

In our imaging experiments a spin echo pulse sequence is used. Three echos are observed and averaged. The first echo is obtained at 12 ms, the second at 28.5 ms, and the third at 43.8 ms. The signal from the short T_2 fraction of sodium is therefore not detected, and only 40% of the sodium signal arising from brain and other tissues is detected. This means that by comparison to protons the abundance of the observable sodium is $\frac{1}{5000}$ that of protons. The clinical sodium images were obtained in a system comprising a 1.5 T magnet operating at a frequency of approximately 16.8 MHz and, as the NMR sensitivity of sodium is 9.25% that of protons at the same magnetic field, the sodium NMR signal obtained from a single echo for a given volume of brain therefore becomes approximately $\frac{1}{54000}$ of the signal obtained from protons at the same field. This weak signal is ameliorated by the following techniques:

1. A faster pulse repetition rate (100 ms) than that used in proton images (0.2–2.0 s). This high repetition rate takes advantage of the relatively short T_1 of sodium.
2. Relaxation of the spatial resolution from 1 mm for protons to approximately 4 mm for sodium.
3. Prolonging the observation time from 4 to 6 min for protons to 34 min for sodium.
4. Using total-volume imaging methods. Although proton imaging is most conveniently carried out as a multislice two-dimensional Fourier method, the use of three-dimensional (3-D) data acquisition techniques allows the observation of sodium at every point of the examined specimen, with each pulse providing additional signal averaging [2]. The acquisition data matrix was originally $32 \times 32 \times 24$ and is now $64 \times 64 \times 24$, covering the entire volume of the human head. In a 34-min study, 24 slices are obtained simultaneously, each one 1 cm thick.

Because of the short T_1 relaxation rate of sodium (55 ms) and the relatively long pulse repetition time (100 ms), there is practically no effect of sodium T_1 relaxation rate on image contrast. The images are, however, strongly T_2-weighted.

Phantom studies in our system demonstrated a contrast sensitivity of 20m*M* of sodium in a disc 4 mm in diameter for a 1-cm-thick slice. This is confirmed by our recent clinical images, where gray/white matter discrimination is seen on scans featuring in-plane resolution of 4 mm.

Optimum clinical evaluation is best achieved by imaging the entire head, if possible. In any 3-D data acquisition scheme, limiting the size of the imaging region by reducing the number of planes observed does not result in a more efficient signal acquisition for a given imaging time and resolution. In the case of sodium imaging in particular, where noise is the limiting factor, a total-volume 3-D imaging approach appears at present to be the most desirable, without much to be gained by limiting the number of observed slices.

In our laboratory a 34-min observation time is used in most cases. We have, however, made 10-min imaging experiments with restless patients, resulting in a corresponding increase in noise. It is interesting to note, however, that in these cases the examination was adequately diagnostic, revealing the lesions with the increased sodium signal.

Results

In Fig. 1, a series of sodium images of the head of a normal volunteer is illustrated. The images are sequential cross-sectional representations extending from the top of the head in the upper left-hand corner down to the level of the eyes in the lower right-hand corner of the illustration. The four sections in the upper row show a bright signal arising from the CSF surrounding the brain in the subarachnoid spaces. The brain parenchyma is the source of little signal and therefore appears darker than the surrounding CSF. The two lateral ventricles and the sylvian fissure are also clearly seen in the second and third rows of images. The suprasellar and perimesencephalic cisterns are seen on the first image of the bottom row. The eyes give a strong signal because of the large extracellular compartment in both vitreous and aqueous chambers. It is important to note that the signal obtained from the normal brain tissue can be readily distinguished from the background air surrounding the

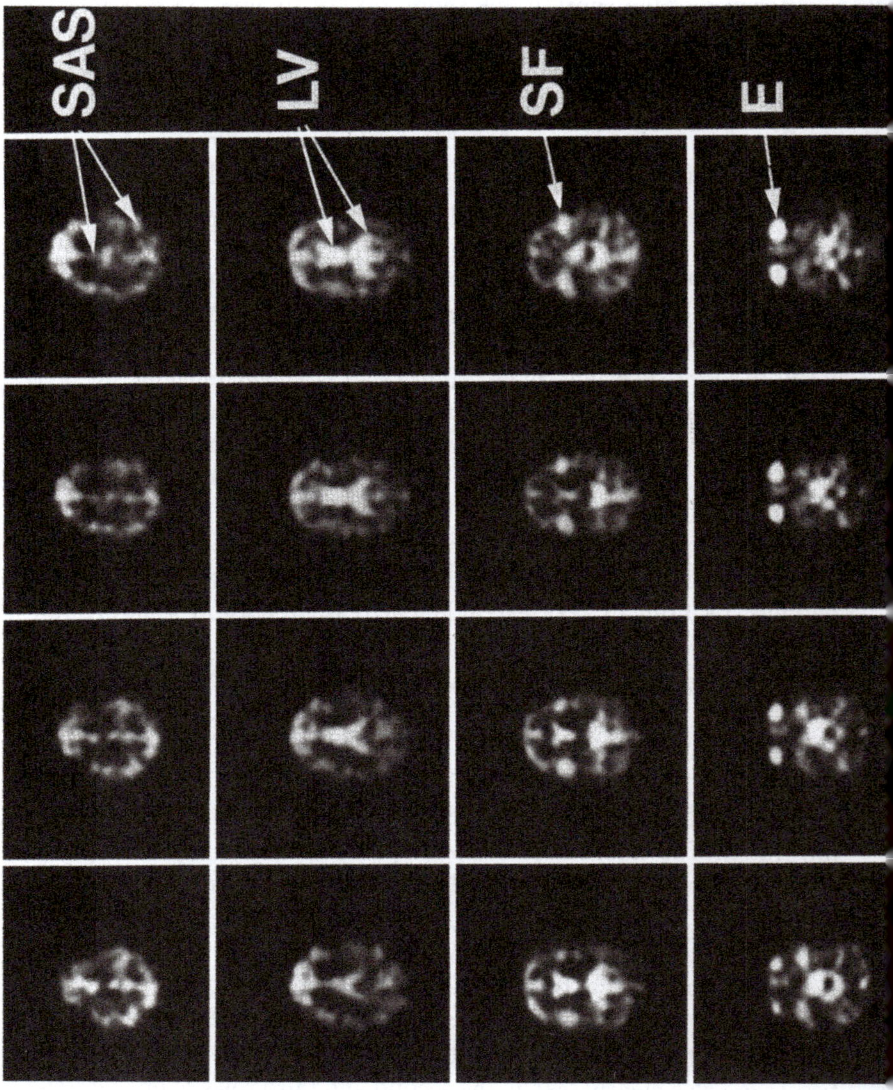

Fig. 1. Serial NMR sodium images obtained in 34 min in a normal volunteer. Only 16 of the 24 sections are illustrated. The data acquisition matrix in this case is 32 × 32 × 24, with in-plane resolution of 6 mm. The images are displayed on a 128 × 128 matrix. The cerebrospinal fluid and the eyes (*E*) are particularly emphasized in these images. *LV,* lateral ventricles; *SAS,* subarachnoid space; *SF,* sylvian fissure

patient's head, indicating the sensitivity of the method to low concentrations of sodium.

Fig. 2 shows a series of sodium images obtained from a normal volunteer using a 64 × 64 × 24 matrix of data acquisition in a 34-min examination. Improvement of the image quality was achieved primarily by the use of ultralow-noise electronics and a high-sensitivity antenna. On the images the differentiation between gray and white matter is beginning to be noticeable. The distinction of the two tissue types is probably due to the difference in size of the vascular bed.

The patient illustrated in Fig. 3 is a hypertensive diabetic with diffuse arteriosclerotic disease. He had a stroke 2 h before the test while in the hospital, resulting in a right hemiparesis. There was also a history of an older stroke from a previous intracranial hemorrhage. The recent stroke is seen as an increased sodium signal in the region of the motor cortex in Fig. 3a, and the older stroke is illustrated in Fig. 3b. Computed tomography (CT) performed with intravenous contrast medium injection immediately after the sodium NMR scan and again a few days later failed to reveal the recent infarction in the motor cortex.

The patient illustrated in Fig. 4 has a vascular malformation seen on the T_2-weighted proton image (Fig. 4a). Around the malformation is a region of high signal, indicating the presence of a region of cerebral damage from a previous hemorrhage. In Fig. 4b the sodium scan shows a greater area of tissue damage extending to and involving the cerebral cortex. In this case the sodium scan showed a greater area of brain damage than was demonstrated on the proton scan.

The patient illustrated in Fig. 5 has a metastatic tumor in the left occipital lobe, which shows an increased sodium signal in a region adjacent to the lateral ventricle. The size of the region of increased sodium signal correlates with the area of cerebral edema seen on CT. On these sodium NMR images there is no distinction between tumor and surrounding edema.

Phantom imaging of sodium chloride solutions in the NMR system used here revealed that the brightness of the image in a given location correlated well with the sodium concentration, in the entire imaging field. Without a reference standard included in the image, one could therefore determine from the local image brightness the relative intensity of the sodium signal arising from the various structures depicted on the same image.

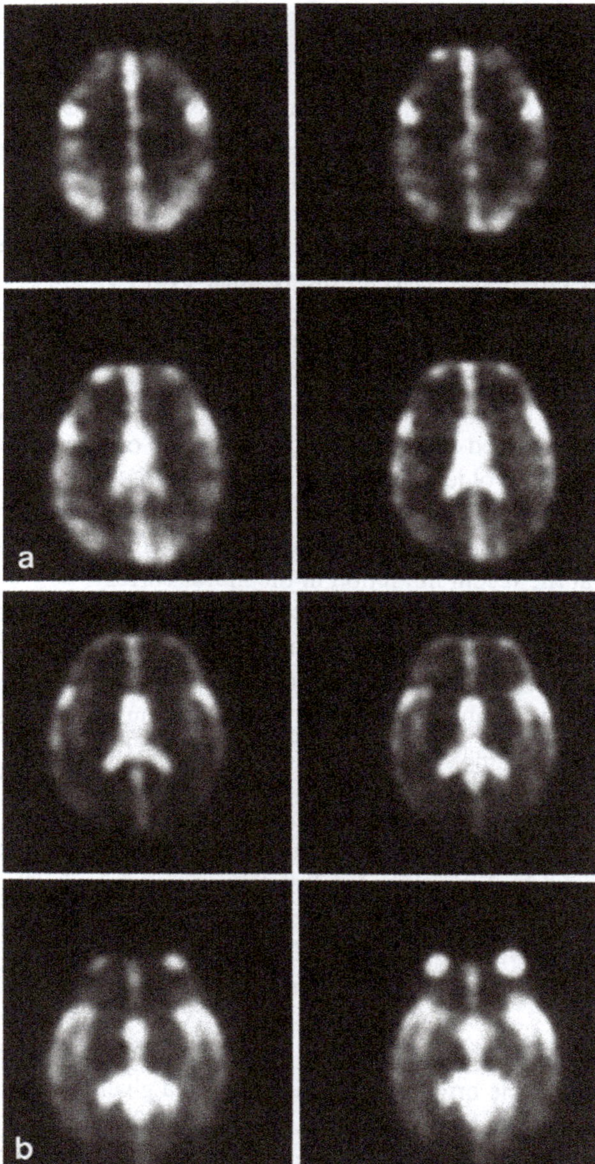
a
b

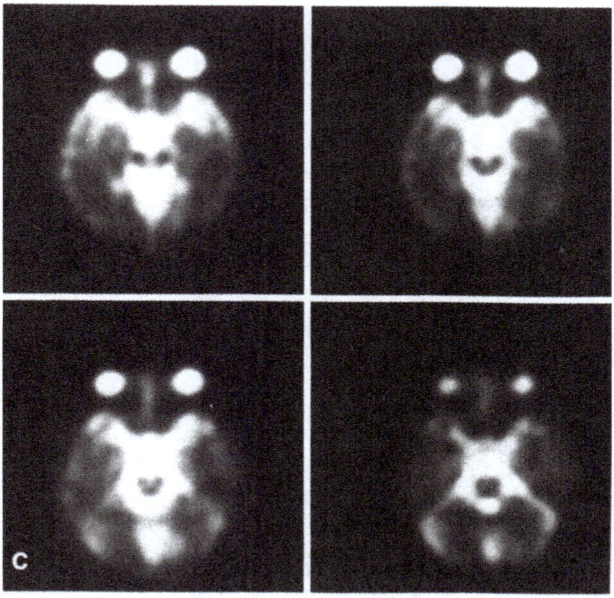

Fig. 2a–c. Recent sodium images of a normal volunteer displayed on a 256 × 256 matrix. Imaging time is identical to that of Fig. 1. The in-plane spatial resolution is 4 mm. Note the gray/white matter discrimination on these images

In the clinical observations, the signal intensity determined from the NMR images of the normal volunteers shows that the lateral ventricles provide a sodium signal four times larger than that obtained from the sourrounding brain parenchyma. Similarly, the old infarction illustrated in Fig. 3b shows a sodium signal four times larger than that obtained from the surrounding normal brain. These values are at variance with direct measurement of sodium concentration obtained by analytical chemical methods. Maximum potential sodium concentration in the region of an infarction should not exceed that in plasma. Sodium is 300% more concentrated in plasma than in brain. The apparent increase in sodium concentration of 400% obtained from the NMR images is presumably due to the reduced NMR visibility of sodium in normal brain when compared with sodium in CSF or in infarction. We conclude that the increased sodium signal observed in the clinical images in stroke is due to both the increased sodium concentration and the increased sodium visibility.

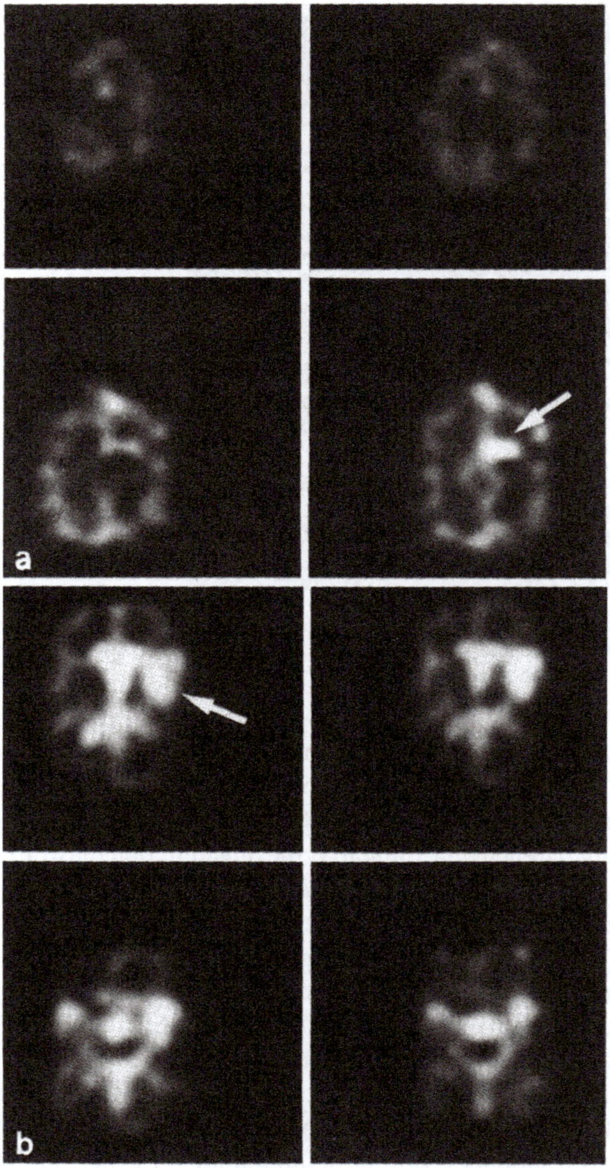

Fig. 3a, b. Sodium NMR images of a patient with a recent and an old stroke. **a** A recent stroke (*arrow*) is shown. This infarction was not seen on CT; **b** The site of the old infarction (*arrow*)

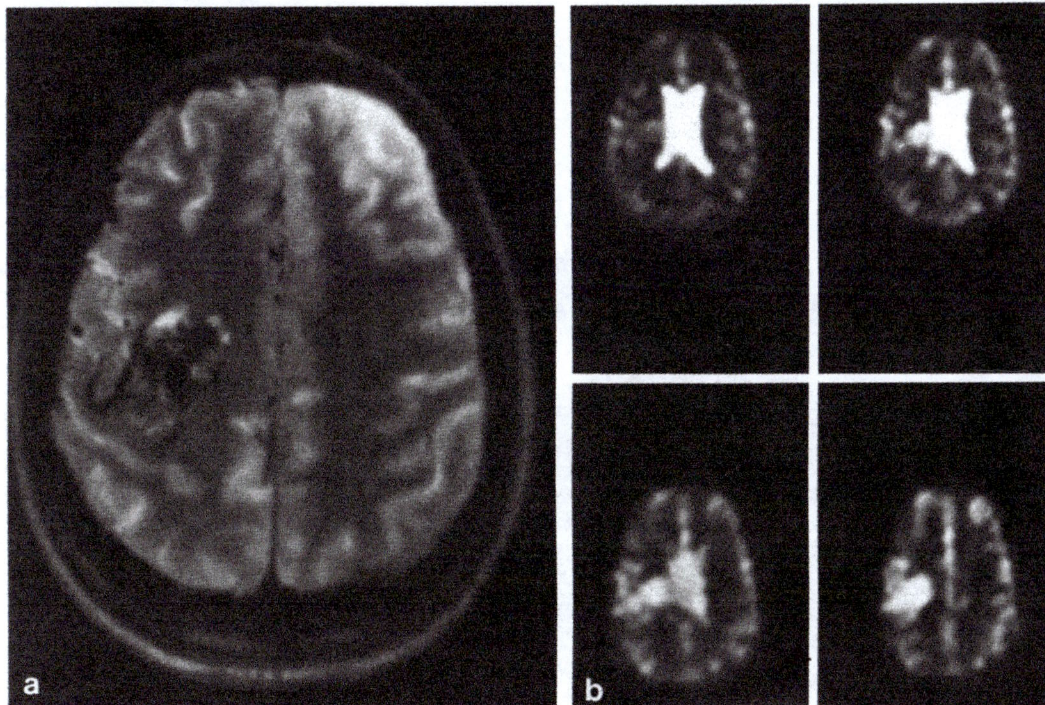

Fig. 4a, b. Proton and sodium NMR scans on a patient with an arteriovenous malformation and a history of a recent hemorrhage and spasm. **a** A T_2-weighted proton image (echo time, 80 ms; repetition time, 4 s). The vascular malformation shows an area of increased signal owing to tissue damage from the recent hemorrhage. The cerebral cortex on the convexity does not reveal any areas of increased proton signal. **b** Sodium scan on the same patient shows the large vascular malformation and the associated area of damage which extends to cover the cerebral cortex on the side of the vascular malformation

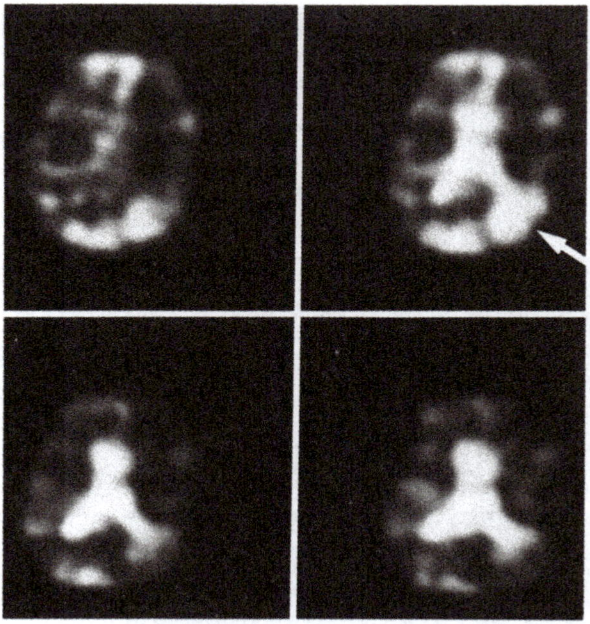

Fig. 5. Sodium NMR scan on a patient with a metastasis. There is a region of increased sodium signal at the site of the tumor and the surrounding edema (*arrow*)

Discussion

Using a currently available 1.5-T magnet and a total-body NMR system developed within this group, we are able to image sodium with a spatial resolution of approximately 4 mm within the plane of study within 34 min.

Our early experience revealed an increased sodium signal in cerebral infarctions, tumors, and cerebral edema. In one case of infarction the lesion was seen on the sodium scan and not on the CT scan, and in a case of arteriovenous malformation of the brain the extent of cerebral damage appeared greater on the sodium than on the proton scans. Recent cases in our experience showed lesions on the sodium scans and not on the T_2-weighted proton images. At present, however, sodium imaging appears to have a high sensitivity for the detection of small or early lesions of the brain. There is evidence that there is increased NMR visibility of sodium in completed infarctions. The possibility of correlating the extent of NMR visibility with the outcome of a stroke should provide an interesting area of research. There is, therefore, an indication that sodium NMR imaging could potentially have greater sensitivity than other methods (CT, proton scanning) for the detection of small or early changes in cerebral disorders.

We believe that with the use of higher field strength magnets, which are already available, even lower-noise electronics, more sensitive antennas, and shorter echo times, it will be possible to achieve a spatial resolution of 2 mm or better or, alternately, to acquire data in a shorter time. In our system, conversion from proton imaging to sodium imaging takes only 15–20 min. For sodium observation, only the radio frequency (RF) antenna and the RF of the spectrometer need to be changed. With this ease of operation, sodium observation may become a standard investigation tool in the clinical environment.

The use of extracellular chemical shift reagent in conjunction with chemical shift imaging could provide additional information on the distribution of sodium between extra- and intracellular spaces in a given region. Preliminary experiments in our laboratory on 10 sedated cats with intrathecal administration of dysprosium tripolyphosphate revealed that this material was tolerated without seizures or any other clinically observable side effects. Intravascular administration of this substance is known to produce serious calcium binding. Investigation is under way for other dysprosium compounds with lesser calcium binding capability. Initial studies using chemical shift imaging of sodium with intrathecal dysprosium administration have revealed the possibility of separately detecting sodium in the extra- and intracellular spaces.

Acknowledgment. This work was supported by National Institutes of Health Grant number CA-28881 and by Philips Medical Systems.

References

1. Berendsen HJC, Edzes HT (1973) The observation and general interpretation of sodium magnetic resonance in biological material. Ann NY Acad Sci 204:459–485
2. Brunner P, Ernst RR (1979) Sensitivity and performance time in NMR imaging. J Magn Reson 33:83–106

3. Cameron IL, Smith NKR, Pool TB, Sparks RL (1980) Intracellular concentration of sodium and other elements as related to mitogenesis and oncogenesis in vivo. Cancer Res 40:1493–1500
4. Civan MM, Shporer M (1978) NMR of sodium-23 and potassium-39 in biological systems. In: Berliner RJ, Reuben J (eds) Biological Magnetic Resonance, vol I. Plenum, New York, pp 1–32
5. Cope FW (1973) Supramolecular biology III. Reinterpretation of NMR data on biological Na^+ in light of the work of Berendsen and Edzes. Ann NY Acad Sci 204:427–431
6. Hilal SK, Maudsley AA, Bonn J, Diffendini R, Simon HE, Mohr J (1984) NMR sodium imaging in stroke: experimental and clinical results (Abstract). Neurology 34 [suppl 1]:260–261
7. Hilal SK, Maudsley AA, Bonn J, Simon HE, Cannon P, Perman WH (1984) NMR imaging of tissue sodium in vivo and in resected organs. Magn Reson Med 1:165–166
8. Hilal SK, Maudsley AA, Simon HE, Perman WH, Bonn J, Mawad ME, Silver AJ, Ganti SR, Sane P, Chien JC (1983) In vivo NMR imaging of tissue sodium in the intact cat before and after acute cerebral stroke. AJNR 4:245–249
9. Hossman KA, Sakaki S, Zimmerman V (1977) Cation activites in reversible ischemia of the cat brain. Stroke 8:77–81
10. Ito U, Ohno K, Nakamura R, Suganuma F, Inaba Y (1979) Brain edema during ischemia and after restoration of blood flow. Measurements of water, sodium, potassium content and plasma protein permeability. Stroke 10:542–547
11. Klatzo I (1967) Neuropathological aspects of brain edema. J Neuropathol Exp Neurol 26:1–14
12. Maudsley AA, Hilal SK (1984) Biological aspects of sodium-23 imaging. Br Med Bull 40:165–167
13. Schuier FJ, Hossman KA (1980) Experimental brain infarcts in cats. Stroke 11:593–600

Uncoupling of Flow and Metabolism in Infarcted Tissue

T. Jones, R. J. S. Wise, R. S. J. Frackowiak, J. M. Gibbs, G. L. Lenzi, and S. Herold

Introduction

In order to study the relationship between regional cerebral blood flow and metabolism, it is necessary to measure these entities of tissue function independently. The use of gamma-ray-emitting radioisotopes to trace physiological/biochemical pathways, and radiation detectors placed external to the body, enables regional tissue function to be measured without disturbing the subject's physiological state. The paper presented by Dr. Ter-Pogossian in this symposium describes how short-lived positron-emitting radioisotopes of oxygen, nitrogen, carbon, and fluorine can be used as labels for in vivo tracer studies. He also demonstrates that by positron emission tomographic (PET) scanning, the transaxial tomographic distribution of these isotopes can be recorded throughout the body. Furthermore, by applying a correction for attenuation effects, it is possible to measure absolute levels of regional tissue concentrations of tracer. This is an important aspect of PET scanning, since it enables concentrations of tracer in blood, measured from samples, to be expressed in the same units as the tracer concentration in the tissues. Thus, appropriate tracer models formulated to define the fate of the labelled molecules can be solved to derive absolute values of the physiological entity being traced.

It is now some 10 years since the introduction of quantitative PET scanning. Since that time, a few tracer procedures have been formulated to measure regional cerebral blood flow (rCBF) and metabolism. To measure blood flow, ^{15}O labelled water [3, 8, 17, 19, 20, 21, 30], ^{77}Kr [39], and ^{18}F-labelled fluoromethane [18] have been used. Oxygen-15, as molecular oxygen, has been used to measure regional cerebral oxygen utilization [3, 8, 26]. Carbon-11 as labelled glucose [6], and both ^{18}F- [27,

31] and ^{11}C-labelled [32] deoxyglucose, have been used to measure glucose transport and utilization in the cerebral tissues. This tracer armamentarium has been exploited at a number of centers to study the relationships between blood flow and metabolism in conditions where cerebral infarction is suspected on clinical and radiological grounds. The purpose of this paper is to summarize the principle findings to date with respect to uncoupling between flow and metabolism. It is confined to reviewing PET studies, while acknowledging that extensive non-PET, single-photon flow measurements in cerebral infarction have been carried out by other workers. In these earlier studies, the researchers have had to infer the relationship between their flow data and tissue metabolism. Combined PET studies enable both components to be studied independently.

Acute Stroke

Although the acute stroke situation is one of the most relevant aspects of cerebral infarction for study, only limited studies have been performed using short-lived cyclotron-produced isotopes and PET scanning. The principle reason for this is the logistical difficulty of adapting a PET program to fit in with the spontaneous presentation of acute stroke victims. These inevitably have to be brought to the PET scanning centers, many of which are not optimally located for acute admissions. The principle findings have been that within the first 24–36 h following the ictus, cerebral tissue in and around the area of depressed flow exhibits a raised oxygen extraction fraction (OEF) [1, 7, 37]. Normally, the cerebral tissues extract 40%–50% of the molecular oxygen per passage through the capillary bed. The finding of a raised oxygen extraction fraction (Figs. 1, 2) demonstrates that the normal balance between metabolic demand for oxygen and blood flow changes to one of ischemia. This indicates that in the acute phase some cerebral tissue within the area of the infarct is surviving by calling upon the normal reserve in oxygen carriage. The measurements of regional cerebral flow and metabolism are somewhat gross, with spatial resolutions of around 1 cm. Therefore, within the lesion examined it is probable that there is a wide heterogeneity of flow and metabolism. A recording of a raised OEF of up to 90% suggests that subpopulations of tissue are completely

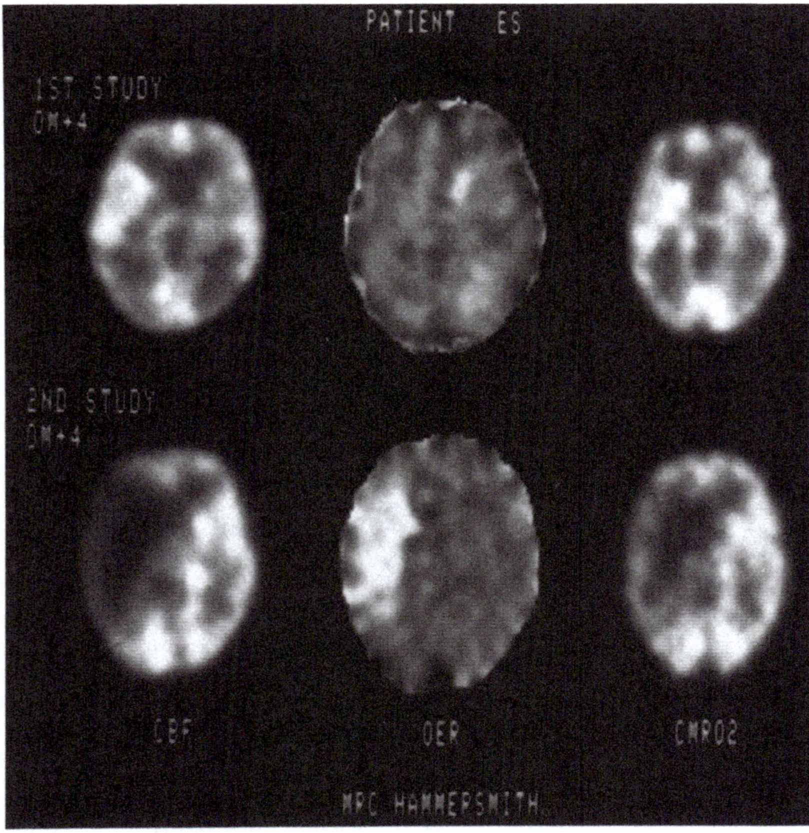

Fig. 1. Regional cerebral blood flow (*CBF*), metabolic rate for oxygen (*$CMRO_2$*), and oxygen extraction ratio (*OER*) in a patient 1 week before (*1st study*) and 7 h after (*2nd study*) a left hemispheric stroke. Although an appreciable reduction in blood flow is seen, the decrease in oxygen utilization is less pronounced. The tissue has compensated by raising the fractional extraction of oxygen seen as a raised OER in the territory affected [37]

exhausting the oxygen available for energy metabolism, and that their rate of oxygen consumption has become flow-rate-dependent. In the face of reduced flow rate, this finding strongly indicates that during the 1st day or so after the stroke, some areas of cerebral tissue remain ischemic in the strict definition of this term. This suggests that an improvement of flow to those areas with high OEF would result in an increase in the rate of oxygen utilization by the tissue.

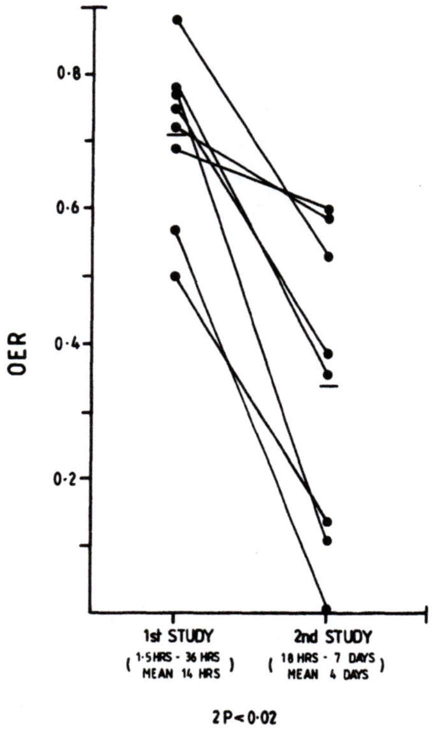

Fig. 2. Quantitative values of the regional oxygen extraction ratio (*OER*) within the territory of acute infarcts in a series of stroke patients. On the 1st day or so following the ictus, OER is raised, indicating the presence of subpopulations of ischemic tissue. Several days later, OER falls to normal or below normal, the latter indicating a state of luxury perfusion [37]

It should be emphasized that in reaching this conclusion, some assumptions have had to be made with respect to data interpretation. The analysis of the recorded ^{15}O metabolic scans centers on a kinetic model which is based on the fact that molecular oxygen, following its utilization, becomes incorporated into water. This involves the donation of hydrogen from the cytochrome system [36]. Scanning procedures using ^{15}O depend on the isolation of the signal component within the cerebral tissue that is due to ^{15}O labelled metabolic water. Using simple, first-order kinetic tracer models, it is possible to derive the rate of oxygen utilization by the tissue, since this determines the amount of ^{15}O incorporated into metabolic water.

There is experimental evidence to indicate that in ischemia the production of labelled water may be uncoupled from the formation of ATP [35]. The extent to which this occurs in the observed human data on oxygen extraction in acute stroke is unknown. To resolve this issue, it would be appropriate to combine measurements in the same stroke patients of regional cerebral oxygen utilization using PET and ATP levels with in vivo nuclear magnetic resonance (NMR) [34].

Stroke — Early Recovery

PET studies in stroke patients carried out around 1 week after the ictus show that the early uncoupling between cerebral flow and metabolism, indicative of ischemia, is reversed. This is demonstrated as a fall in the fractional extraction of oxygen to or below the normal value (Figs. 1, 2) [1, 7, 37]. This reversal is brought about by a return of blood flow to the cerebral tissue within and peripheral to the infarct as a result of recannulation of previously occluded vessels, opening up of collateral channels, or new vessel formation. Furthermore, in quantitative terms, the demand of the tissue for oxygen falls, adding to the reversal in tissue oxygen extraction [37]. The reflow rCBF has sometimes been reported to be higher than normal (Fig. 3), and these PET data are compatible with those of Lassen, who reported the phenomenon of luxury perfusion in stroke patients using the ^{133}Xe method [22]. This pathologic coupling of flow to metabolism to produce a low value for fractional extraction in the tissue is in line with the early reports of Feindel et al., who described "red veins" which drained the area of the cerebral infarct [10]. There is often comment that this low fractional oxygen extraction is due to arteriovenous shunting in and around the lesion. However, the flow method employed depends on measurement of the uptake and retention of diffusible inert tracers where exchange with the tissue parenchyma occurs at the capillary level. It has also been shown that the ratio between oxygen and glucose utilization is altered within the infarcted area of the recovering stroke patient [38] (Figs. 3, 4). In the presence of a low fractional extraction of oxygen, glucose utilization was found to be increased in comparison with that in surrounding normal tissue. It was concluded that this metabolic uncoupling is due to increased glycolytic activity within the lesion, owing to the presence of infiltrating white cells. Support for this interpretation comes from earlier histological studies, where increased populations of white cells in cerebral infarction have been clearly demonstrated [40].

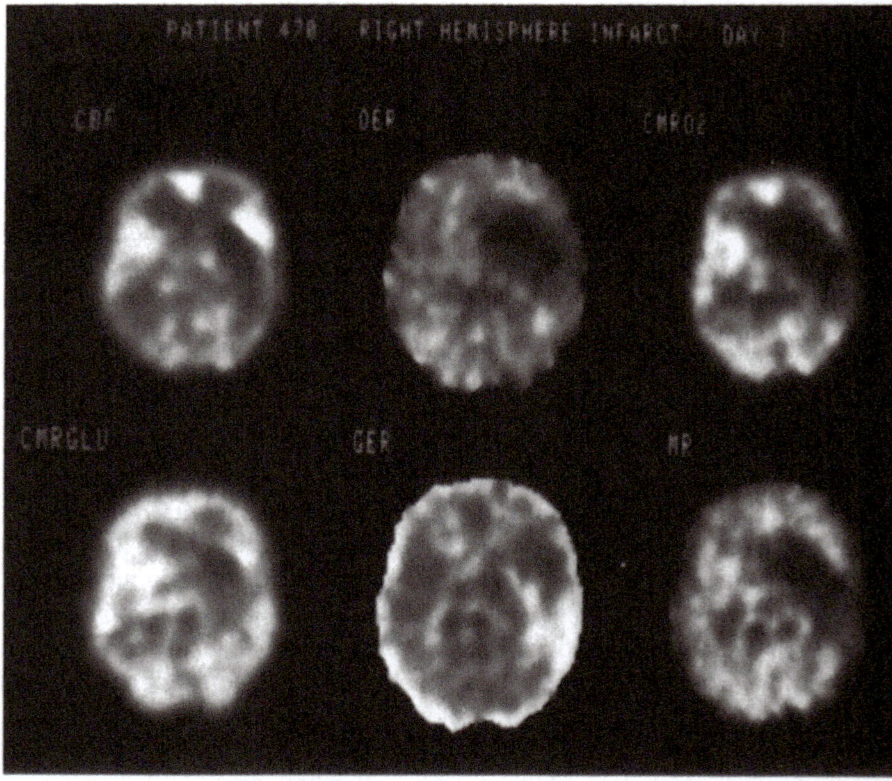

Fig. 3. Regional cerebral blood flow (*CBF*), and metabolic rate for oxygen (*CMRO*$_2$) and for glucose (*CMRGl*) in a right hemispheric infarct, 3 days following the ictus. A state of luxury perfusion exists within the lesion, as demonstrated by the focal regions of high CBF coupled with low $CMRO_2$ and low oxygen extraction ratio (*OER*). The glucose utilization in the territory is higher than that of oxygen. This is clearly demonstrated as an area of low metabolic ratio (r*MR*) — oxygen utilization divided by glucose utilization. Although the coupling between CBF and $CMRO_2$ is changed within the region (*OER*), that between CBF and CMRGl, the glucose extraction ratio (*GER*), is fairly well preserved [38]

Chronic Infarction

PET measurements in patients more than a month or so after the ictus show that blood flow and oxidative metabolism are reduced in and around the lesion, while the coupling between the two reverts to the balance found in normal cerebral tissues [1, 7, 23]. This residual focal

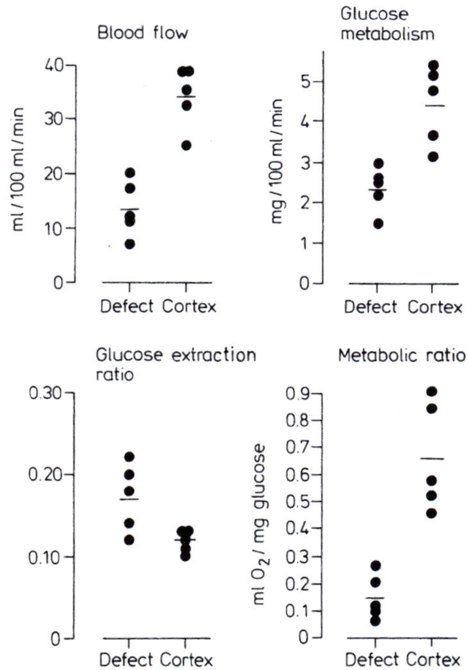

Fig. 4. Quantitative values of blood flow, oxygen metabolism, glucose metabolism and the ratio of these: metabolic ratio (*MR*) for territories within established infarcts and the contralateral cortex [38]. The low MR within the lesion is indicative of glycolytic activity, which could be explained by the infiltration of white cells [40]. The glucose extraction ratio (GER, glucose uptake to blood flow), tends to be normal or high in the area of the lesion

deficit has been found to be associated with metabolic and flow reductions in the contralateral cerebellum and hemisphere [2, 14, 23, 24, 33]. The paper by Pawlik in this symposium deals specifically with the remote effects of focal lesions. It is clear from the literature that there are insufficient data to correlate the sequential changes in regional cerebral flow and metabolism with the neurologic outcome and recovery following a stroke. More in-depth studies are needed to determine the role such changes play in a patient's functional recovery and in the degree of the residual deficit.

Studies of chronic multi-infarct dementia patients have shown focal reductions in cerebral oxygen and glucose utilization, which have a normal coupling to tissue blood flow [5, 9]. Further evidence that flow meets regional tissue metabolic demand comes from systematic PET studies of the effect of a vasodilator drug in multi-infarct patients. Gibbs et al. [11] have shown that measurements before and after a 6-month course of medication demonstrate no change in the coupling between blood flow and oxygen utilization. Also, PET studies of chronic stroke patients prior to and following omentum transposition to revascularize tissue peripheral to the infarct have demonstrated no change in tissue perfusion or coupling with oxygen utilization [15].

Pre-infarction

There is considerable current interest in using PET to study patients who are categorized as being at risk from a cerebral infarct as a result of arterial occlusion. In particular, patients with a history of transient ischemic attacks (TIAs) associated with carotid artery disease have been studied. The $rCMRO_2$ tends to be lower than that of age-matched controls [12]. This finding, which increases with the degree of arterial occlusion (Fig. 5), suggests that neurologically silent microinfarctions occur. In the more advanced cases, the tissues exhibit an increase in the fractional extraction of oxygen (Figs. 5, 6), thus demonstrating a change in the normal coupling between flow and metabolic demand [4, 12, 25, 29]. This suggests a progressive fall in oxygen reserve for the cerebral tissues as the degree of arterial occlusion increases. This state of critical or misery perfusion has been reported to be relieved by bypass surgery [4]. Further evidence for this fall in perfusion pressure is demonstrated by the observation that prior to calling upon the reserve in oxygen carriage, the cerebral vasculature is seen to dilate in order to reduce

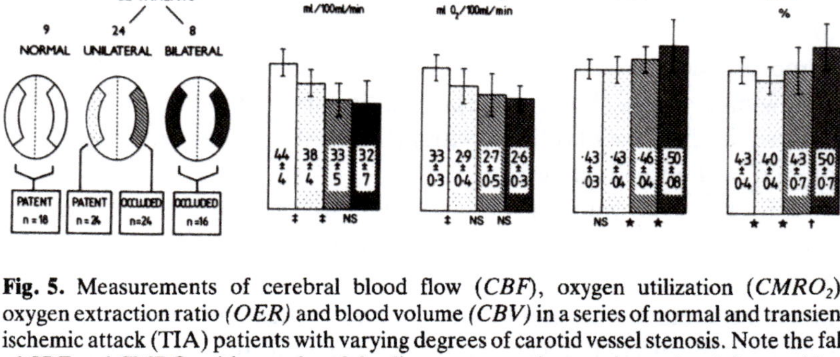

Fig. 5. Measurements of cerebral blood flow (*CBF*), oxygen utilization (*$CMRO_2$*), oxygen extraction ratio *(OER)* and blood volume *(CBV)* in a series of normal and transient ischemic attack (TIA) patients with varying degrees of carotid vessel stenosis. Note the fall of CBF and $CMRO_2$ with severity of the disease, suggesting previous microinfarcts. Also, OER and CBV increase with severity, indicating the loss of perfusion reserve and pressure with advancing disease [12]. *NS,* not significant

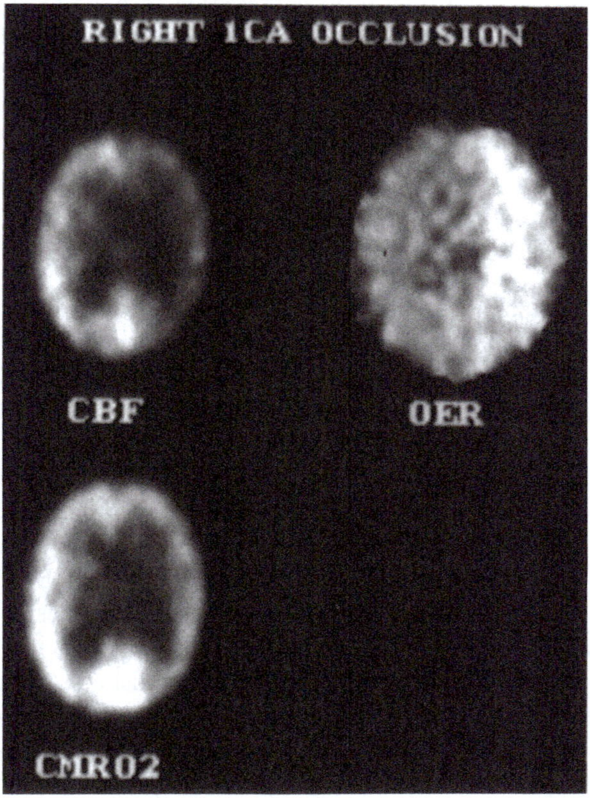

Fig. 6. Images of CBF, $CMRO_2$ and OER in a patient with advanced carotid stenosis. Note the raised OER on the *right side,* compensating for the reduced perfusion reserve. Also, the level of $CMRO_2$ is still decreased in that hemisphere

peripheral resistance (Fig. 5). While the fractional extraction of oxygen remains normal, the ratio of regional cerebral blood volume to blood flow is seen to increase [12, 28], but this has been reported to revert to normal following vascular surgery [13, 29].

Sickle cell anemia patients, another clinical category of patients in whom cerebral infarction tends to occur, have been studied with PET. They have an incidence of major strokes at a young age. A PET study of young adults has shown that regional cerebral oxygen utilization is mismatched to blood flow. Although the fractional oxygen extraction is normal because of the low oxygen carriage capacity of blood in this disease, the flow itself is increased to maintain the required level of oxygen delivery per unit of time. It is also of interest to note that the levels of oxygen utilization and blood flow in the brain tissue of these patients are slightly reduced, indicating that a series of microinfarcts may have occurred [16].

Discussion and Summary

Although a relatively limited number of PET studies have been performed in cerebral infarct patients, especially in the acute phase, some new information has emerged from the research to date. The finding of a raised OEF in and around the infarct in the first 24 h suggests that there may be tissues within that territory which could be salvaged if the flow were increased or the metabolic demand of the tissue decreased. Before introducing measures to accomplish such changes, there is a need to ensure that the raised extraction of oxygen is in fact coupled with the production of ATP. A combined PET and in vivo NMR spectroscopic study should provide a means to clarify this position. The relevance of this early uncoupling, indicative of ischemia, to the eventual clinical neurologic outcome has yet to be established. A large series of systematic serial studies will be necessary to determine the longitudinal changes that underlie good or poor clinical recovery. Such a program requires commitment by PET groups suitably located near centers geared to receiving and treating acute stroke patients. As PET scanners with higher spatial resolution are introduced, it will be possible to look more closely at the location of the raised oxygen extraction — the ischemic territory. The finding that oxygen and glucose utilization become uncoupled in an established infarct points to the need for caution when using tracers of glucose incorporation in this pathology. The understanding is that once white cells begin to infiltrate the area, their high glycolytic rate gives a distorted view of the metabolism of the cerebral tissues themselves. In those patients at risk from cerebral infarcts who present with TIAs, uncoupling between flow and oxygen utilization indicates a trend toward chronic ischemia. This marker is beginning to prove highly suggestive of the need for vascular surgery. There is also a suggestion from the data reported to date that the cerebral tissues of recovered TIA patients have a lower metabolic rate for oxygen coupled with lower blood flow. This indicates that such patients have experienced microinfarction. In patients with long-standing multi-infarct dementia, the surviving tissue is normally coupled, although reduced regional cerebral metabolic rates and blood flow have been found.

This brief review has attempted to itemize the principle findings with PET to date with respect to uncoupling between tissue metabolism and flow in cerebral infarcts. A number of extended reviews have been written on the results found in cerebral infarction, and a recommended reading list of these reviews is included in the reference list. In addition, an extended reading list is given of the PET methods developed to measure regional cerebral blood flow and metabolism, together with books which deal with the use of PET in cerebral studies in general.

References

1. Ackerman RH, Correia JA, Alpert NM, Baron JC, Gouliamos A, Grotta JC, Brownell GL, Taveras JM (1981) Positron imaging in ischemia stroke disease using compounds labelled with oxygen-15. Arch Neurol 38:537–543
2. Baron JC, Bousser MG, Comar D, Castaigne P (1981) Crossed cerebellar diaschisis in human supratentorial brain infarction. Trans Am Neurol Assoc 105:459–461
3. Baron JC, Bousser MG, Comar D, Soussaline F, Castaigne P (1981) Non-invasive tomographic study of cerebral blood flow and oxygen metabolism in vivo: potentials, limitations and clinical applications. Eur Neurol 20:273–284
4. Baron JC, Bousser MG, Rey A, Guillard A, Comar D, Castaigne P (1981) Reversal of local "misery-perfusion syndrome" by extra-intracranial arterial bypass in hemodynamic cerebral ischemia: a case study with ^{15}O positron tomography. Stroke 12:454–459
5. Benson DF, Kuhl DE, Hawkins RA, Phelps ME, Cummings JL, Tsai SY (1983) The fluorodeoxyglucose ^{18}F scan in Alzheimer's disease and multi-infarct dementia. Arch Neurol 40:711–14
6. Blomqvist G, Bergstrom K, Bergstrom M et al. (1984) Models for ^{11}C glucose. In: Greitz T et al. (eds) The metabolism of the human brain studied with PET. Raven, New York, pp 185–194
7. Bousser MG, Baron JC, Iba-Zizen MT, Comar D, Cabans E, Castaigne P (1980) Migrainous cerebral infarction: a tomographic study of cerebral blood flow and oxygen extraction fraction with the oxygen-15 inhalation technique. Stroke 11:145–148
8. Frackowiak RSJ, Lenzi GL, Jones T, Heather JD (1980) Quantitative measurement of regional cerebral blood flow and oxygen metabolism in man using ^{15}O and positron emission tomography: theory, procedure and normal values. J Comput Assist Tomogr 4(6):727–736
9. Frackowiak RSJ, Pozzilli C, Legg NJ, Du Boulay GH, Marshall J, Lenzi GL, Jones T (1981) Regional cerebral oxygen supply and utilisation in dementia: A clinical and physiological study with oxygen-15 and positron tomography. Brain 104:753–778
10. Feindel W, Perat P (1965) Red cerebral veins, a report on arteriovenous shunts in tumours and cerebral scans. J Neurosurg 22:315–325

11. Gibbs JM (1985) Cerebral blood flow and metabolism in dementia, with reference to the effects of pharmacological intervention. In: Trimble M (ed) Proceedings of British Association of Psychopharmacology meeting, October 1984. Pergamon, Oxford
12. Gibbs JM, Wise RJS, Leenders KL, Jones T (1984) Evaluation of cerebral perfusion reserve in patients with carotid-artery occlusion. Lancet I:310–314
13. Gibbs JM, Wise RJS, Ross Russell RW, Mansfield AO (1985) Cerebral circulatory reserve before and after surgery for occlusive carotid artery disease. Procceedings of the twelfth international symposium on cerebral blood flow and metabolism, June 1985. J Cereb Blood Flow Metab [Suppl] In press
14. Heiss WD, Pawlik G, Wagner R, Ilsen HW, Herholz K, Wienhard K (1983) Functional hypometabolism of noninfarcted brain regions in ischemic stroke. J Cereb Blood Flow Metab 3 [suppl 1]:582–583
15. Herold S, et al. (In preparation) Regional brain blood flow and oxygen utilisation pre and post omental transposition in established cerebral infarction
16. Herold S, Brozovic M, Gibbs JM, Lammertsma AA, Leenders KL, Carr D, Fleming JS Measurement of regional cerebral blood flow, blood volume and oxygen utilisation in patients with sickle cell anaemia. (in preparation)
17. Herscovitch P, Markham J, Raichle ME (1983) Brain blood flow measured with intravenous $H_2{}^{15}O$. I. Theory and error analysis. J Nucl Med 24:782–789
18. Holden JE, Gatley SJ, Hichwa RD, Ip WR, Shaughnessy WJ, Nickles RJ, Polcyn RE (1981) Cerebral blood flow using PET measurements of fluoromethane kinetics. J Nucl Med 22:1084
19. Huang SC, Carson RE, Phelps ME (1982) Measurement of local cerebral blood flow and distribution volume with short-lived isotopes: general input technique. J Cereb Blood Flow Metab 2:99
20. Huang SC, Carson RE, Hoffman EJ, Carson J, MacDonald N, Barrio JR, Phelps ME (1983) Quantitative measurement of local cerebral blood flow in humans by positron computed tomography and ^{15}O-water. J Cereb Blood Flow Metab 3:141
21. Kanno I, Lammertsma AA, Heather JD, Gibbs JM, Rhodes CG, Clark JC, Jones T (1984) Measurement of cerebral blood flow using bolus inhalation of $C^{15}0_2$ continuous inhalation method. J Cereb Blood Flow Metab 4:224–234
22. Lassen NA (1966) The luxury-perfusion syndrome and its possible relation to acute metabolic acidosis localized within the brain. Lancet II:1113–1115
23. Lenzi GL, Frackowiak RSJ, Jones T (1982) Cerebral oxygen metabolism and blood flow in human cerebral ischemic infarction. J Cereb Blood Flow Metab 2:321–335
24. Martin WRW, Raichle ME (1983) Cerebellar blood flow and metabolism in cerebral hemisphere infarction. Ann Neurol 14:168–176
25. Martin WRW, Baker RP, Herscovitch P, Zeiger HE, Grubb RL, Raichle ME (1983) The selection of patients for extracranial-intracranial bypass surgery: hemodynamic and metabolic criteria. Neurology (NY) 32:A89
26. Mintun MA, Raichle ME, Martin WRW, Herscovitch P (1984) Brain oxygen utilization measured with 0-15 radiotracers and positron emission tomography, J Nucl Med 25:177–187
27. Phelps ME, Huang SC, Hoffman EJ, Selin C, Sokoloff L, Kuhl DE (1979) Tomographic measurement of local cerebral glucose metabolic rate in humans with (F-18)-2-fluoro-2-deoxy-D-glucose: validation of method. Ann Neurol 6:371
28. Powers W, Martin W, Herscovitch P, Raichle M, Grubb R (1983) The value of regional cerebral blood volume measurements in the diagnosis of cerebral ischemia. J Cereb Blood Flow Metab 3 [Suppl 1]:598–599

29. Powers WJ, Martin WRW, Herscovitch P, Raichle ME, Grubb RL (1984) Extracranial-intracranial bypass surgery: hemodynamic and metabolic results. Neurology (in press)
30. Raichle ME, Martin WRW, Herscovitch P, Mintun MA, Markham J (1983) Brain blood flow measured with intravenous $H_2^{15}O$. II. Implementation and validation. J Nucl Med 24:790–798
31. Reivich M, Kuhl D, Wolf A, Greenberg J, Phelps ME, Ido T, Casella V, Fowler J, Hoffman EJ, Alavi A, Som P, Sokoloff L (1979) The (^{18}F)-fluorodeoxyglucose method for the measurement of local cerebral glucose utilization in man. Circ Res 44:127
32. Reivich M, Alavi A, Wolf A, Greenberg JH, Fowler J, Christman D, MacGregor R, Jones SC, London J, Shiue C, Yonekura Y (1982) Use of 2-deoxy-D-[1-^{11}C] glucose for the determination of local cerebral glucose metabolism in humans: variation within and between subjects. J Cereb Blood Flow Metab 2:307
33. Rougemont D, Baron JC, Lebrun-Grandie P, Bousser MG, Cabanis E, Laplane D (1983) Debnit sanguin cerebral et extraction d'oxygène dans les hémiplégies lacunaires. Etude semi-quantitative par l'oxygène 15 et la tomographie d'émission. Rev Neurol 139 (4):277–282
34. Shulman GI, Alger JR, Prichard JW, Shulman RG (1984) Nuclear magnetic resonance spectroscopy in diagnostic and investigative medicine. J Clin Invest 74:1127–1131
35. Siesjö BK (1981) Cell damage to the brain: a speculative synthesis. J Cereb Blood Flow Metab 1:155–185
36. Ter-Pogossian MM, Eichling JO, Davis DO, Welch MJ (1970) The measure in vivo of regional cerebral oxygen utilization by means of oxyhaemoglobin labelled with radioactive oxygen-15. J Clin Invest 49:381
37. Wise RJS, Bernardi S, Frackowiak RSJ, Legg NJ, Jones T (1983) Serial observations on the pathophysiology of acute stroke: the transition from ischaemia to infarction as reflected in regional oxygen extraction. Brain 106:197–222
38. Wise RJS, Rhodes CG, Gibbs JM, Hatazawa J, Palmer T, Frackowiak RSJ, Jones T (1984) Disturbance of oxidative metabolism of glucose in recent human cerebral infarcts. Ann Neurol 14:627–637
39. Yamamoto YL, Thompson CJ, Meyer E, Robertson JS, Feindel W (1977) Dynamic positron emission tomography for study of cerebral haemodynamics in cross section of the head using positron-emitting ^{68}Ga-EDTA and ^{77}Kr. J Comput Assist Tomogr 1:43
40. Yates PO, Blackwood W, Carselius JAN (eds) (1976) Vascular diseases of the nervous system. Greenfields neuropathology, 3rd edn. Arnold, London, pp 86–147

Further Reading

Cerebrovascular Disease and PET

Frackowiak RSJ (1982) The positron emission scanner: application to clinical neurology and the understanding of cerebral ischaemia. In: Barnett HJM (ed) AAN annual course no 203: cerebrovascular disease. American Academy of Neurology pp 151–172
Frackowiak RSJ, Wise RJS (1983) Positron tomography in ischemic cerebrovascular disease. Neurologic Clinics 1 (1):183–200

Frackowiak RSJ, Wise RJS, Gibbs JM, Jones T (1984) Positron emission tomographic studies in aging and cerebrovascular disease at Hammersmith Hospital. Ann Neurol 15 [Suppl]:112–118

Kuhl DE, Phelps ME, Howell AP, Metter EJ, Selin C, Winter J (1980) Effects of stroke on local cerebral metabolism and perfusion: mapping by emission computed tomography of ^{18}FDG and $^{13}NH_3$. Ann Neurol 8:47–60

Powers WJ, Raichle ME (1985) Positron emission tomography and its application to the study of cerebral vascular disease in man. Stroke (to be published)

Wise RJS (1982) Regional cerebral blood flow and oxygen metabolism in acute stroke. In: Sarner M (ed) Advanced medicine 18. Pitman Medical, London, pp 302–311

Wise RJS, Gibbs JM, Frackowiak RSJ (1985) The application of PET to the study of cerebral perfusion reserve and ischaemia. In: Greitz T, Ingvar DH, Widén L (eds) The metabolism of the human brain studied with positron emission tomography. Raven, New York, pp 363–375

PET Methods to Measure Cerebral Blood Flow and Metabolism

Ell PJ, Holman BL (eds) (1982) Computed emission tomography. Oxford University Press, Oxford, pp 188–210

Greitz T et al. (eds) (1984) The metabolism of the human brain studied with positron emission tomography. Raven, New York

Heiss WD, Phelps ME (eds) (1983) Positron emission tomography of the brain. Springer, Berlin Heidelberg New York

Jones T, Frackowiak R, Lenzi GL (1983) Clinical application of measurement of regional cerebral blood flow and oxygen utilization by positron emission tomography. In: Heiss WD, Phelps ME (eds) Positron emission tomography of the brain. Springer, Berlin Heidelberg New York, pp 107–112

Kuhl DE (1984) Imaging local brain function with emission computed tomography. Radiology 105 (3):625–631

Lammertsma AA, Jones T (1983) The correction for the presence of intravascular oxygen-15 in the steady state technique for measuring regional oxygen extraction ratio in the brain. 1. Description of the method. J Cereb Blood Flow Metab 3:416–424

Lammertsma AA, Jones T, Frackowiak RSJ, Lenzi GL (1981) A theoretical study of the steady state model for measuring regional cerebral blood flow and oxygen utilization using oxygen-15. J Comput Assist Tomogr 5(4):544–550

Lammertsma AA, Heather JD, Jones T, Frackowiak RSJ, Lenzi GL (1982) A statistical study of the steady state technique for measuring regional cerebral blood flow and oxygen utilization using oxygen-15. J Comput Assist Tomogr 6:566–573

Lammertsma AA, Wise RJS, Heather JD, Gibbs JM, Leenders KL, Frackowiak RSJ, Rhodes CG, Jones T (1983) The correction for the presence of intravascular oxygen-15 in the steady state technique for measuring regional oxygen extraction ratio in the brain. 2. Results in normal subjects and brain tumour and stroke patients. J Cereb Blood Flow Metab 3:425–431

Lebrun-Grandie P, Baron JC, Soussaline F, Loch'h C, Sastra J, Bousser MG (1983) Coupling between regional blood flow and oxygen utilization in the normal human brain. A study with positron tomography and oxygen 15. Arch Neurol 40:230–236

Leenders KL, Gibbs JM, Frackowiak RSJ, Lammertsma AA, Jones T (1984) Positron emission tomography of the brain: new possibilities for the investigation of human cerebral pathophysiology. Prog Neurobiol 23:1–38

Pantano P, Baron JC, Lebrun-Grandie P, Duquesnoy N, Bousser MG (1984) Regional cerebral blood flow and oxygen consumption in human aging. Stroke 15(4):635–641
Phelps ME, Mazziotta JC, Huang SC (1982) Study of cerebral function with positron emission tomography. J Cereb Blood Flow Metab 2:113–162
Reivich M (ed) (1985) Positron emission tomography. Liss, New York
Research issues in positron emission tomography. Proceedings of a conference sponsored by the National Institute of Neurological and Communications Disorders and Stroke, 1983. Ann Neurol 15 [Suppl] (1984)
Rhodes CG, Lenzi GL, Frackowiak RSJ, Jones T, Pozzilli C (1981) Measurement of CBF and $CMRO_2$ using the continuous inhalation of $C^{15}O_2$ and $^{15}O_2$. J Neurol Sci 50:381–389

Remote Effects of Focal Lesions on Cerebral Flow and Metabolism

G. PAWLIK, K. HERHOLZ, C. BEIL, R. WAGNER, K. WIENHARD and W.-D. HEISS

Impairment of regional energy metabolism and hemodynamics in morphologically unaltered areas of the brain beyond the bounds of ischemic tissue damage is a common finding in positron emission tomography (PET) of stroke patients. This transneural depression of function resulting from an actual brain lesion was first described by von Monakow [38]. He used the term "diaschisis" to denote the phenomenon of transient dysfunction, e.g., in the cerebral hemisphere contralateral to a cortical infarct, which may initially be widespread, but finally and gradually through unknown repair processes returns to whatever level it will naturally attain. Physiological evidence in support of this concept was presented by Kempinsky [17], who demonstrated that for up to 1 h following unilateral cortical ablation there is depressed electrical activity at homotopic points in the opposite hemisphere, and by Hoedt-Rasmussen and Skinhoj [14], who found a bilateral decrease in cerebral blood flow (CBF) after ischemic stroke. The latter observation was confirmed in several other two-dimensional CBF studies [25, 27, 36], but a more precise demonstration of such regional effects occurring far from the site of infarction was feasible only after the development of PET and appropriate tracer kinetic methods for noninvasive studies of local cerebral metabolism and blood flow [9, 16, 30, 31, 33, 37].

Asymmetries are highly conspicuous, even on transaxial tomographic images showing only relative activity distributions. Therefore, asymmetric deactivation of various gray matter structures ipsilateral to a supratentorial lesion [10, 19, 32], and of the contralateral cerebellum [1, 2, 21, 29], was among the first PET findings reported on stroke or tumor patients. With increasing methodological sophistication, however, some functional depression was also found on the contralateral cerebral and ipsilateral cerebellar hemisphere [6, 11, 20, 22, 24]. Meanwhile, it is generally accepted that remote deactivation is a variable phenomenon occurring when neuronal units have been deprived of the established input of facilitating impulses which they normally receive from now disabled or disconnected source units within the central nervous system (CNS). However, in view of the abundance of known anatomical pathways, the complex nature of neuronal interaction, and the comparatively small and inhomogeneous groups of patients so far investigated for diaschisis, the ongoing controversy among PET workers as to the anatomical and pathophysiological determinants of certain topographical patterns of functional deactivation is hardly surprising. In particular, the location and other characteristics of a supratentorial lesion predisposing to contralateral cerbellar hypometabolism and blood flow have been at issue [2, 20, 22, 24, 29].

This study was designed to shed some light on the stochastic aspects of functional anatomy pertinent to the problem of diaschisis in ischemic stroke, by the application of advanced analytical strategies to the metabolic and hemodynamic PET data obtained from a larger series of patients in comparison with healthy control subjects. However, because of the extremely large — at present unavailable — sample sizes required for multivariate analyses, the nature of this work still remains exploratory; interpretation of results must be limited to major syndromes and can only be tentative.

Subjects

Control Group

Thirty normal subjects (23 men, 7 women), whose ages ranged from 16.0 to 61.6 years with a median age of 33.55 years, in good general physical condition and without mental or neurologic abnormalities on clinical examination, underwent both a dynamic and a static PET examination as outlined below. The informed consent of all control persons was obtained.

Patients

Sixty-two patients (31 men, 31 women), ranging in age from 18.4 to 79.3 years with a median age of 61.25 years, who had suffered their first ischemic stroke between 3 days and 31 years (median interval 17.5 days) previously, were subjected to the same PET procedure as the controls for assessment of their brain glucose metabolism. Of those patients, 14 had a PET flow measurement 2–3 h prior to their metabolic study. In all cases, the diagnosis of completed stroke was confirmed by the clinical course and by x-ray computed tomography (CT). No patient had more than one focal ischemic lesion, but all had a distinct neurologic deficit ranging from mild aphasia/dysarthria to hemiplegia/ataxia and stupor. Of the 62 patients, 38 had a left, and 24 a right hemispheric lesion, with 34 infarcts involving the cerebral cortex, 9 the thalamus, 7 the brainstem, 6 the striatum, 3 the perinuclear white matter (internal capsule and cerebral peduncle), and 3 being located in either of the cerebellar hemispheres. Infarct size as estimated planimetrically from the CT images ranged from 0.18–294.69 ccm, with a median volume of 13.90 ccm. The smallest ischemic lesions were usually old thalamic cysts, while all of the larger infarctions were cerebral hemispheric insults including part of the cortex.

Methods

Metabolic PET Studies

The general procedure of PET scanning has been described in detail elsewhere [12]. Approximately 15 min before the beginning of the PET examination, after the subject had been familiarized with the laboratory environment to minimize tension and anxiety, short Teflon catheters were inserted into an antecubital vein for injection and into a dorsal vein of the contralateral hand that was subsequently kept in a temperature-controlled water bath at 44 °C for blood sampling according to a previously described schedule [40]. Subjects were then comfortably placed on a reclining chair with their head inside the gantry of a four-ring Scanditronix PC 384 positron camera [8], requested to keep their eyes closed until the end of the study, and given a rapid intravenous bolus injection of ~5 mCi 2[^{18}F]fluoro-2-deoxy-D-glucose (FDG) in 5 ml sterile, nonpyrogenic, normal saline solution. Concurrently, recordings were started: seven equally spaced, 11-mm-thick slices centered from the canthomeatal line to 82 mm above it were simultaneously scanned for 40 min, at consecutive intervals gradually increasing from 1–5 min. A continuous write-out of the instantaneous whole-head coincidence count rate was obtained on paper chart for monitoring the stability of head positioning. Ambient noise and activities of personnel in the dimly lit PET room were minimal. This procedure yielded reliable dynamic information on the tracer accumulation in virtually all major brain structures [40]. Subsequently, only the normal control subjects were displaced relative to the scanner along the craniocaudal axis by 6.9 or 20.6 mm, and scanning continued for another 10 min, so that the images from the dynamic and from the latter static recording period were intercalated by half a center-to-center distance. Data from the tomographic device and plasma count rates from a cross-calibrated well counter, as well as plasma glucose values determined in duplicate by a standard enzymatic method, were stored in the memory of a VAX 11/780 (DEC) computer for later processing. The spatial activity distribution in the image plane was reconstructed using an edge-finding algorithm to determine the skull contour for attenuation correction [4], a deconvolution for subtraction of scattered radiation [5], and a filtered backprojection algorithm resulting in a 128 × 128 matrix that was displayed by means of a RAMTEK 9050 system as an interpolated 256 × 256 pixel image. All measured radioactivities were corrected for decay before they were entered in any model equation. Activity images recorded 30–40 min and 40–50 min post injection were transformed into quantitative metabolic maps according to the $k_{1,3}$-normalization procedure [40].

Hemodynamic PET Studies

Local cerebral blood flow (lCBF) was measured using the [^{18}F]-methylfluoride (MF) method proposed and validated by Holden et al. [15]. PET examinations were performed under ambient conditions similar to the dynamic metabolic studies, except for the following: a short catheter for frequent blood sampling was inserted into a lower arm vein, the canthomeatal line as indicated by a laser beam was marked on the patient's head with permanent ink to allow precise repositioning for the later FDG run, and a respirator mask with a three-way stopcock for the control of the ventilation gas was tightly fitted to the patient's nose and mouth. Approximately 40 mCi of MF, produced according to the method of Wagner [39], was administered to the patient by his rebreathing for 2 min from a spirometer equipped with a soda lime CO_2 absorber and an oxymeter controlling the oxygen supply to the closed system to keep the oxygen concentration in the inhaled air constant at 21%. This was followed by 25 min of room air ventilation. Throughout the procedure, a constant flow of ~300 ml/min of gas was extracted from a sampling port on the respirator mask and pulled in double-walled, heated tubing through an NE-102 plastic scintillation detector and an infrared CO_2 gas analyzer; finally it was returned to the circle system or issued to a gas trap. A custom-built micro-computer recorded the MF activity of the sampled gas at a frequency of 100 channels/min. At the first MF inhalation, a sequence of seventeen 40-s PET scans was initiated, and blood samples were drawn into preweighed, sealed syringes at gradually increasing intervals and counted in the cross-calibrated sample changer. Following the reconstruction of serial activity images according to the principles outlined above, the lCBF values and local tissue/blood partition coefficients on the seven scanned brain slices were estimated pixel by pixel from the local brain clearance curves. The arterial MF input function derived from the end-tidal expired gas curve and a scaling factor was determined by fitting a multicompartment model to the series of venous activities as described by Koeppe et al. [18].

Regionalization

For topography-related statistical analyses of the physiological PET data, a comprehensive set of regions of interest (ROI) was marked on all functional images by means of an interactive FORTRAN program in connection with the computer's image display system, providing fast and largely user-independent mapping with high reproducibility and flexible adaption to individual anatomical variations [13]. Briefly, after the outer brain contour had been determined on a tomographic image by edge finding, generous raw region contours were automatically placed relative to that outer contour one after the other, according to geometrical standard parameters specified in a permanent file. For example, for the cerebral cortex, the process was conducted in angular steps of 20° measured from the center of the brain contour. Within each of those raw regions, the final subregion of interest was marked so as to cover all pixels with an activity above a pre-defined proportionate level (gray matter). A more complex definition applied to white matter. The contours of ischemic lesions as demonstrated by CT were manually copied onto the PET images by joystick input, and were excluded from the above regionalization procedure. That way, a whole set of seven slices was represented by approximately 150 subregions comprising separate infarct, gray, and white matter regions, for each of which the mean metabolic and flow rate was computed individually. Corresponding $k_{1,3}$-normalized metabolic values were, however, exclusively employed in the clustering procedure described below. Otherwise, because of the well-known shortcomings of any autoradiographic model in pathologic tissue, only metabolic rates calculated from regional rate constants determined by standardized nonlinear curve-fitting [40] were used in both patients and normal controls.

For optimum anatomical comparability, at first the average set of subregions characteristic of each brain level along the craniocaudal axis was established in the normal subjects and divided into an anterior, middle, and posterior segment. In order to compensate for differences among individuals with respect to the shape of the base of the brain and the height of the cerebrum, all the possible combinations of upward and downward shifts by at most one step (13.7 mm) of each subject's slice segments were then tested for correspondence between the region names of the individual and reference sets, utilizing the Tanimoto coefficient as the measure of similarity. The best-fitting segmental level adjustment then became the basis of further analysis.

Considerable data reduction was achieved, and the variability of numerical results minimized, by subjecting the regional metabolic rates from all 14 slices scanned in the control group to a pattern recognition procedure with topographical boundary conditions: first, subregions were grouped into an optimum number of clusters according to Ward's method (SAS procedure Cluster, SAS Institute Inc, Cary, North Carolina); then, anatomically neighboring subregions belonging to the same metabolic cluster were pooled to from a larger, topographically meaningful region. For each of the 21 regions thus defined on either hemisphere — Fig. 1 shows a lateral projection — the average flow (rCBF) and glucose metabolic rates (rCMRGl), weighted by the size of the constituting subregions, were calculated from the subregion data. Similarly, global measures were obtained by pooling either all cerebral hemispheric or all noninfarcted brain regions, including white matter.

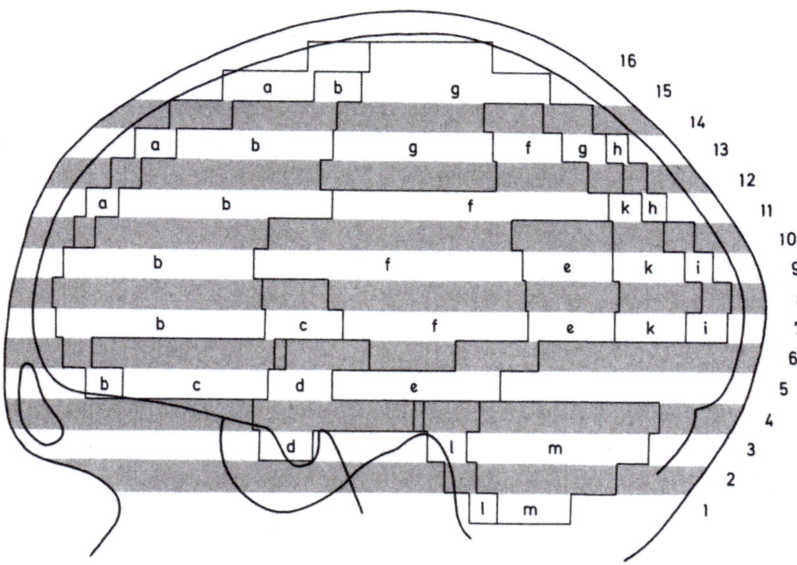

a) Prefrontal cortex
b) Midfrontal cortex
c) Inferior frontal cortex
d) Temporopolar cortex
e) Posterior temporal cortex
f) Temporal parietal cortex
g) Parietal cortex
h) Visual median cortex
i) Occipitopolar cortex
k) Occipitotemporal cortex
l) Inferior brainstem
m) Cerebellum

Fig. 1. Standardized lateral skull projection with tomographic slice positions and topometabolic cluster regions determined in 30 normal human subjects. *Shaded areas* indicate common levels of dynamic PET scanning

Statistical Analysis

All numerical variables were tested for goodness of fit to a normal distribution with parameters estimated from the sample, using a modified Kolmogorov-Smirnov test [23]. Parametric methods were applied to normally distributed data only. Whenever data were nongaussian, summary statistics are reported as median and range, otherwise as arithmetic mean ± standard deviation (SD). In all correlation analyses, Spearman's rho (ϱ) was computed to allow comparison of differently skewed variables. *P* values of multiple comparison tests, adjusted according to the Bonferroni method, are indicated by subscripts. Contingency tables were analyzed by exact procedures, and linear modelling of remote effects was performed by multiple regression (SAS procedure REG, SAS Institute Inc, Cary, North Carolina).

Results and Discussion

General

Although the patients were significantly older than the normal subjects ($P << 0.00001$, by the Wilcoxon-Mann-Whitney test), the latter could still serve as controls because in neither group were global (ϱ: patients −.146, controls −.136), hemispheric (ϱ: patients −.189, controls −.132), or regional metabolic rates (ϱ: patients −.211 to .030, controls −.128 to .114) notably correlated with age. This observation is in good agreement with the findings of other investigators, e.g., by Duara et al. [7] in a series of 40 healthy men of all ages — no matter how much absolute levels of CMRGl, depending largely on technical details, differ among laboratories [12]. Reference values of rCMRGl from the present control series, except for the larger white matter regions, are listed in Table 1.

In patients, all the metabolic rates in noninfarcted brain regions were highly intercorrelated, with coefficients ranging between .432 (thalamus/inferior frontal cortex, $P_{adj} < 0.01$) and .930 (posterior temporal/temporal parietal cortex, $P_{adj} < 0.0001$) (adj, adjusted). Compared with the differentiated correlation structure found by Metter et al. [26] in healthy adults, this relative homogeneity may be indicative of

Table 1: rCMRGl in 30 normal subjects

Region	rCMRGl (μmol/100 g/min) (mean ± SD)	Percentage difference between sides $\left[\frac{\text{(left-right)} \times 100}{\text{Max. (left, right)}}\right]$ (mean ± SD)
Whole brain	34.0 ± 5.08	−1.10 ± 2.148
Prefrontal cortex	41.5 ± 6.79	.87 ± 5.168
Midfrontal cortex	43.0 ± 7.84	− .23 ± 3.600
Inferior frontal cortex	35.0 ± 5.67	1.21 ± 8.505
Insular cortex	39.2 ± 8.53	−2.90 ± 10.891
Temporopolar cortex	31.4 ± 5.61	1.72 ± 8.326
Posterior temporal cortex	36.2 ± 5.76	−3.67 ± 5.865
Temporoparietal cortex	39.3 ± 6.68	−3.04 ± 4.017
Parietal cortex	38.6 ± 5.98	−1.91 ± 4.416
Visual and medial cortex	42.7 ± 7.31	.02 ± 4.214
Occipitopolar cortex	35.3 ± 7.18	1.88 ± 7.782
Occipitotemporal cortex	31.8 ± 4.76	−2.84 ± 4.055
Striatum	40.8 ± 7.43	.73 ± 7.321
Thalamus	37.1 ± 7.66	.06 ± 6.800
Perinuclear white matter	19.7 ± 3.21	− .46 ± 8.760
Superior brainstem	28.3 ± 4.09	3.53 ± 10.862
Inferior brainstem	24.3 ± 4.09	−4.20 ± 10.267
Cerebellum	29.8 ± 4.71	−1.76 ± 4.493

some pathological synchronization among all regions of a focally injured brain. However, no such universal interrelation was detected among regional flow rates. Correlations were significant only among superior frontal (ϱ: .779, $P_{adj} < 0.05$), temporal (ϱ: .834 to .908 $P_{adj} < 0.01$), and occipital areas (ϱ: .913 to .971, $P_{adj} < 0.01$), and subcortical gray matter regions (ϱ: .879, $P_{adj} < 0.01$), as well as between infratentorial structures and superior frontal cortex (ϱ: .801, $P_{adj} < 0.05$), parietotemporal and inferior frontal cortex (ϱ: .808, $P_{adj} < 0.01$), parietal cortex and subcortical gray matter (ϱ: .801, $P_{adj} < 0.05$), and superior brainstem and striatum (ϱ: .774, $P_{adj} < 0.05$). Some of those relationships may be explained by a common arterial supply, others by true functional interaction. Anyhow, rCMRGl would appear to be the more specific indicator of neuronal activity.

Both regional metabolism and blood flow within the ischemic lesion exhibited some trend toward lower values at larger infarct sizes, but because of considerable data scatter that correlation was significant only in the larger sample of rCMRGl (ϱ: $-.319$, $P < 0.05$). This phenomenon must be attributed in part to the relatively poor spatial resolution common to all nuclear medicine imaging techniques; in part it may be due to more favorable supply conditions in the border zone of smaller infarctions [28].

Diffuse Metabolic Effects

When rCMRGl in the apparently unaffected cerebral and cerebellar hemisphere (i.e., ipsi- and contralateral to a cerebellar infarct, or contra- and ipsilateral to an infarct of any other location) was compared with the respective metabolic rates in the control group (Table 2), patients with ischemic lesions centered in the midfrontal, inferior

Table 2. Effect of lesion sites on "unaffected" cerebral and cerebellar hemisphere metabolism

Localization of Lesion	n	rCMRGl (µmol/100 g/min) (mean ± SD) on "unaffected" cerebral hemisphere	cerebellar hemisphere
Control	30	35.8 ± 4.83	33.3 ± 5.13
Prefrontal cortex	2	34.4 ± 6.43	31.8 ± 1.27
*Midfrontal cortex	9	30.8 ± 6.77	29.3 ± 5.11
*Inferior frontal cortex	6	29.1 ± 2.84	31.0 ± 4.63
Insular cortex	2	31.4 ± 6.15	28.0 ± 5.66
Temporopolar cortex	1	26.8	25.4
Posterior temporal cortex	1	25.3	21.5
***Temporoporietal cortex	10	27.1 ± 5.25	27.3 ± 6.54
Partietal cortex	1	19.8	19.6
Occipitotemporal cortex	2	33.7 ± 0	32.0 ± 4.03
Striatum	6	31.1 ± 6.07	29.2 ± 4.65
*Thalamus	9	32.4 ± 3.52	28.2 ± 3.63
Perinuclear white matter	3	31.2 ± 3.70	29.1 ± 3.27
***Superior brainstem	3	22.7 ± 1.82	18.1 ± 1.29
**Inferior brainstem	4	28.6 ± 3.45	23.9 ± 9.36
**Cerebellum	3	24.8 ± 3.30	24.6 ± 1.31

Significance of difference from control group: P_{adj}, * < 0.05, ** < 0.001, *** < 0.00001 ($_{adj}$, adjusted)

frontal, temporoparietal, and thalamic or infratentorial regions showed significant functional deactivation of those remote areas, as opposed to patients with infarcts in the striatum or perinuclear white matter. This effect was nonselective, i. e., both target regions were affected either in similar proportion or not at all. Figure 2 demonstrates such a diffuse metabolic depression without corresponding flow reduction in a patient who had suffered a left inferior brainstem infarction and was alert at the time of study, 3 days after the onset of symptoms. Similarly diffuse transhemispheric and posterior fossa depression was also suggested by the PET data obtained by Lenzi et al. [22] in 14 patients with supratentorial infarction, particularly when consciousness was impaired.

Martin and Raichle [24] also noted a generalized effect on the cerebellum related to frontal lobe and parietooccipital infarction in 13 stroke cases. Likewise, Kushner et al. [20] found significantly lower absolute metabolic rates in both cerebellar hemispheres of a group of 16 patients with ischemic or tumorous cerebral lesions. However, neither of those reports mentioned any dependence of diffuse deactivation on the site of infarction, while the present observation of a differential influence is in line with clinical experience indicating that certain infarcts, e. g., of the internal capsule, are not commonly accompanied by global functional impairment, such as organic brain syndrome or disturbance of consciousness.

Although it was not a purpose of this study to give a characterization of the evolution of diaschisis, which would require a prospective longitudinal design, even the cross-sectional data available may provide some valuable information pertaining to that problem, particularly since the patients analyzed represent a random sample from all stroke patients studied by PET in this laboratory. Despite substantial variation among individuals, a significant upward trend with time was found for rCMRGl in the "unaffected" cerebral hemisphere (ϱ, .347; $P < 0.01$), asymptotically approaching the range of normality approximately 3 months after the ictus. A similar trend was apparent for blood flow, but even larger data scatter precluded statistical significance in the comparatively small sample. The recovery of global brain function thus suggested may be the physiological basis of the improvement in the activities-of-daily-living score often seen even in stroke victims showing little change in their specific neurologic deficit.

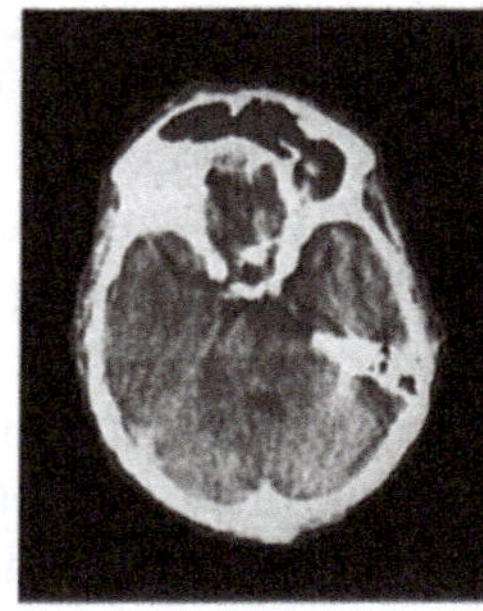

Fig. 2. CT (*left*), CBF (*Fig. 26*), and CMRGl (*Fig. 2c*) images of a 43-year-old man with left inferior brainstem infarction. All images are oriented so that anterior is up and right-side structures are on the viewer's left

Asymmetrical Deactivation

Metabolic asymmetries were tested by comparing the percentage difference between homotopic regions on either side (Table 1), individually for each patient, to the respective 95% prediction interval of the control group. Differences exceeding those limits were declared operationally significant. According to that definition, only 4 of the 62 patients (6.5%; 3 thalamic, 1 cerebellar infarction) exhibited no metabolic asymmetry other than the lesion, and 91.5% of all asymmetrically depressed regions were located ipsilateral to the infarct and in the contralateral cerebellar hemisphere. Among all the possible target regions of remote effects, the cerebellum has attracted the most attention, perhaps because it is both clearly at a distance from any supratentorial focus and is also one of the few anatomically well-defined gray matter structures most frequently affected by functional deactivation. Nevertheless, in the literature, opinions on the location of a lesion predominantly causing cerebellar metabolic asymmetry are quite contradictory: Lenzi et al. [22] implicated the parietal lobe, Martin and Raichle [24] the frontal lobe, Kushner et al. [20] the parietal lobe, and Patronas et al. [29] primarily the frontal lobe again. Those discrepancies may at least in part be explained by topographical ambiguity and sample bias owing to nonproportional representation. Therefore, in the present series, cerebral cortical infarctions were grouped into a prerolandic and a postrolandic category and compared with respect to asymmetrical deactivation of deeper gray matter structures (Table 3). Both infarct size ($P > 0.10$, by the Wilcoxon-Mann-Whitney test) and rCMRGl of the ischemic lesion ($P > 0.10$, by one-way analysis of variance) were similar in the two

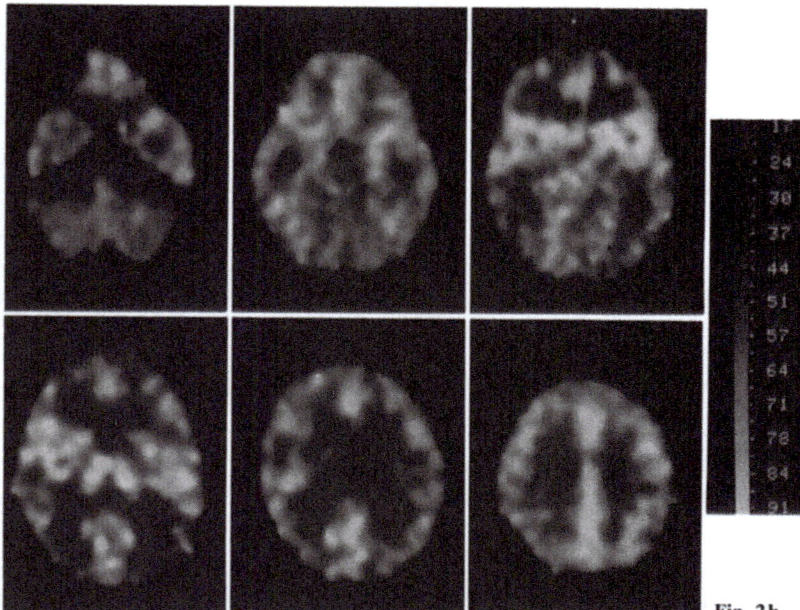

Fig. 2b

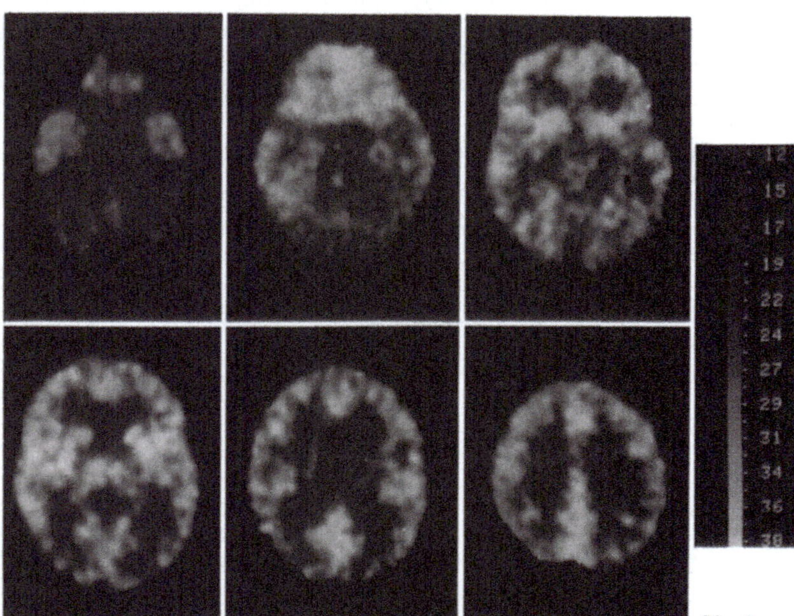

Fig. 2c

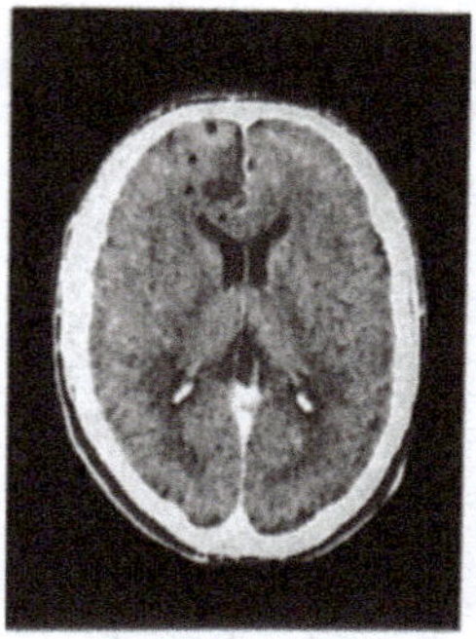

Fig. 3. CT (*left*), CBF (*Fig. 3b*), and CMRGl (*Fig. 3c*) images of a 30-year-old man with right prefrontal infarction. All images are oriented so that anterior is up and right-side structures are on the viewer's left

groups. Neither for the multivariate set ($P > 0.10$, by the Freeman-Halton test) nor for the cerebellum alone ($P > 0.10$, by Fisher's exact test) was the observed heterogeneity statistically significant. Usually, ipsilateral basal ganglia and thalamus and contralateral cerebellum had a common response as might have been expected considering the anatomical multiloop connections among those structures. Figure 3 demonstrates this pattern of diaschisis in a patient with an infarction in the right anterior cerebral artery territory, 5 months after onset of symptoms, showing only mild asymmetrical, remote effects that are proportionate for blood flow and metabolism. Figure 4, by contrast, exemplifies a predominantly hemodynamic pattern of widespread asymmtrical func-

Table 3. Metabolic asymmetry effect of cerebral cortical lesions on basal ganglia, thalamus, and cerebellar cortex

Asymmetrical deactivation of			Localization of cerebral cortical lesion					
ipsilateral		contra-lateral						
Striatum	Thalamus	Cerebellum	prerolandic *n*	%	postrolandic *n*	%	total *n*	%
yes	yes	yes	10	52.6	4	26.7	14	41.2
no	no	no	3	15.8	3	20.0	6	17.6
no	yes	no	2	10.5	4	26.7	6	17.6
no	no	yes	1	5.3	3	20.0	4	11.8
yes	yes	no	2	10.5	0	0	2	5.9
no	yes	yes	0	0	1	6.7	1	2.9
yes	no	yes	1	5.3	0	0	1	2.9
yes	no	no	0	0	0	0	0	0
			19	100	15	100	34	100

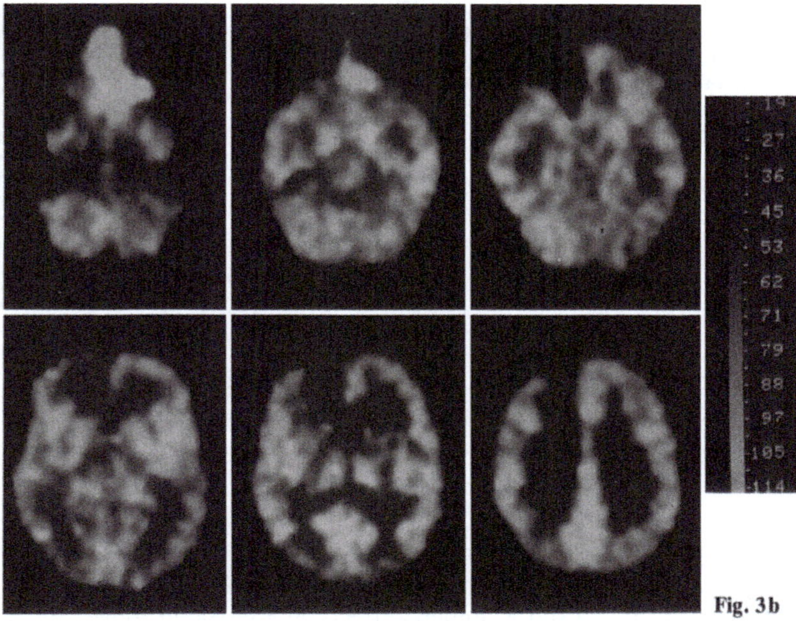

Fig. 3b

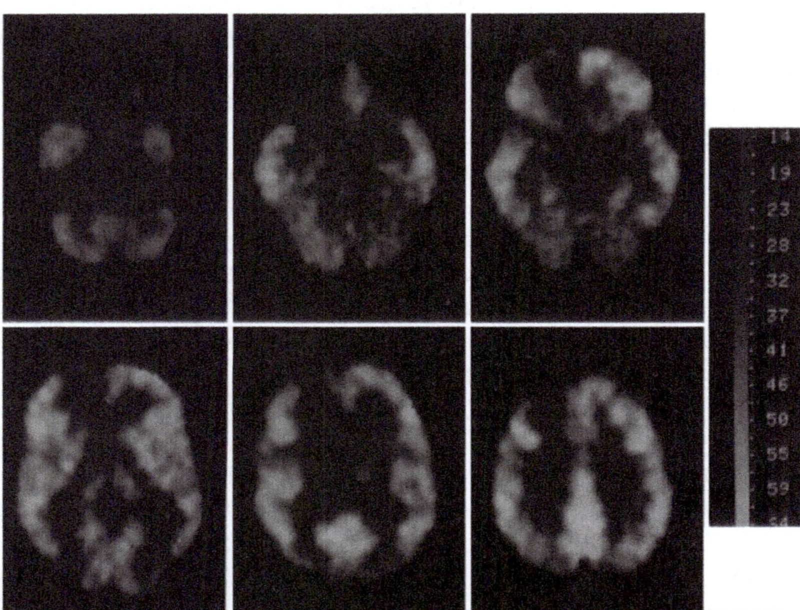

Fig. 3c

Fig. 4. CT (*left*), CBF (*Fig. 4b*), and CMRGl (*Fig. 4c*) images of a 70-year-old man with right temporoparietal infarction. All images are oriented so that anterior is up and right-side structures are on the viewer's left

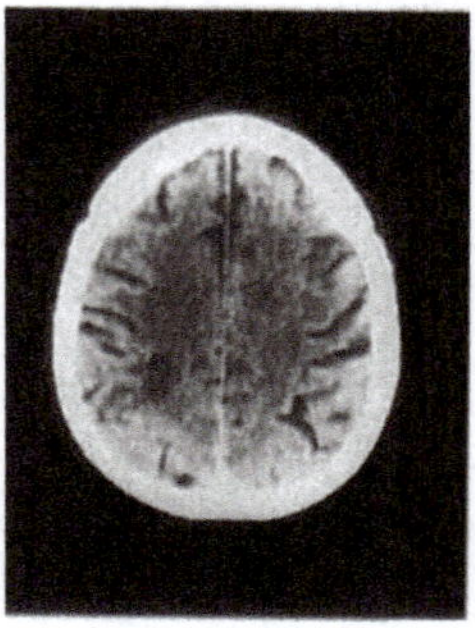

tional deactivation equally involving the ipsilateral cortex and deeper gray matter, in a patient with an infarction centered in the right temporoparietal region, 40 days after his stroke.

Modelling of Metabolic Remote Effects

In view of the global and asymmetry effects demonstrated above, several simple linear hypotheses were tested for their goodness of fit to the patient data. Quite surprisingly, a very small number of variables proved to be sufficient to explain most of the observed variance: the metabolic rate in any remote region ipsilateral to a supratentorial infarction, or in the contralateral cerebellum, was best described in terms of the sum of the metabolic rates in the contralateral cerebral hemisphere and in the infarct region, respectively, weighted by factors depending on the location of the lesion. The finding of a significant relationship between lesion metabolic rate and the activity of other brain regions is in good agreement with results of animal experiments demonstrating a close correlation of the neurologic deficit and the adenosine triphosphate (ATP) contents of an ischemic lesion [34].

As can be seen in Table 4, the fit was excellent ($P < 0.001$) for all the combinations of noninfarcted regions and lesion sites represented in large enough numbers, with coefficients of determination (proportion of variance explained) ranging between .885 and .999, and a median of 99%. However, the standardized regression coefficients reflecting the relative import of the two independent variables did not in every case differ significantly from zero. This means that direct and indirect effects

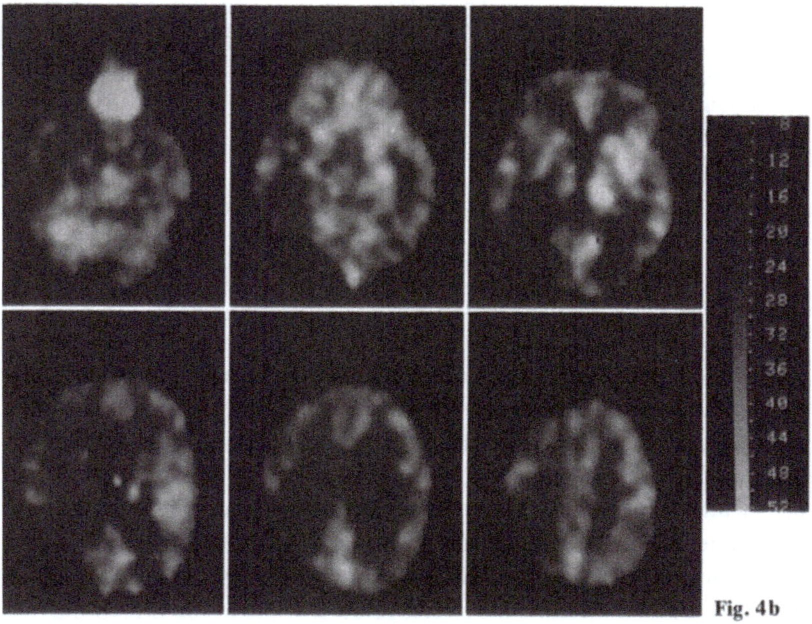

Fig. 4b

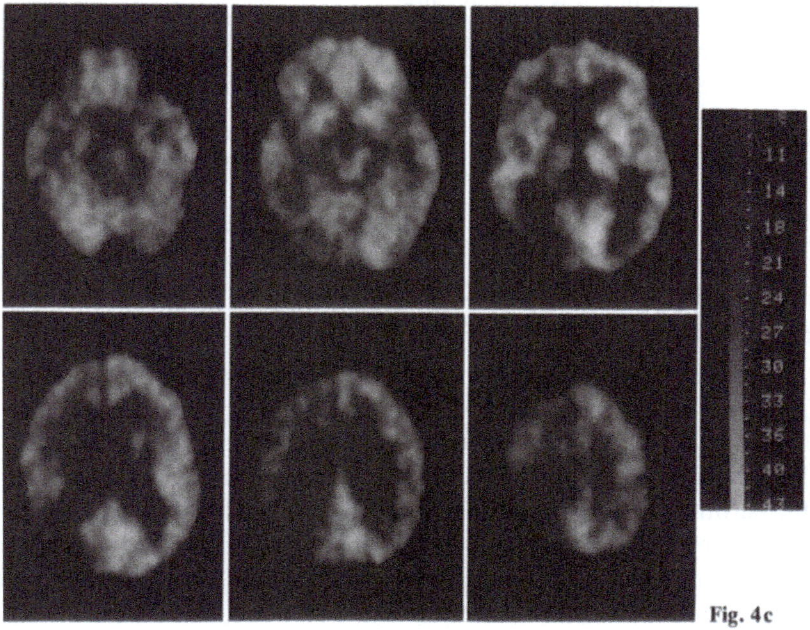

Fig. 4c

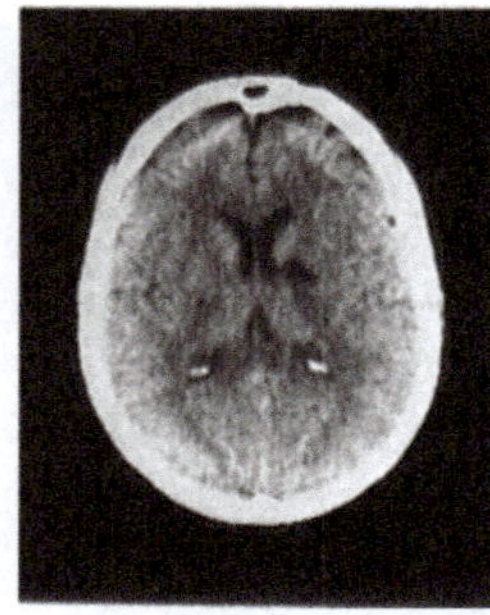

Fig. 5. CT (*left*), CBF (*Fig. 5b*), and CMRGl (*Fig. 5c*) images of a 49-year-old women with an infarction of the left striatum, demonstrating almost proportionate, diffuse hemodynamic and metabolic depression of all, but predominantly ipsilateral, brain structures, and contralateral cerebellum, 8 days after onset of symptoms. All images are oriented so that anterior is up and the patient's right is on the viewer's left

contributed in variable proportion to the total metabolic effect on a certain region in the respective group of patients characterized by the same infarct localization. While midfrontal and temporoparietal infarctions and ischemic lesions in the striatum and thalamus produced significant diffuse effects on almost all other ipsilateral regions and the contralateral cerebellum (Fig. 5), the only systematic remote effects caused by inferior frontal infarcts were unspecific, focussed primarily on the occipital lobe. Significant effects related to the lesion metabolic rate were detected in a few regions only:

1. The prefrontal cortex in thalamic infarction
2. The temporopolar cortex in lesions of the midfrontal cortex or striatum
3. The posterior temporal cortex in temporoparietal infarction
4. The parietal cortex in thalamic infarction
5. The occipitotemporal cortex in midfrontal infarction
6. The thalamus in temporoparietal infarction
7. The superior brainstem in thalamic lesions
8. The contralateral cerebellum in midfrontal and in thalamic infarction.

The latter direct effects are easily attributed to well-known connecting fiber tracts, while indirect effects must be ascribed to the functional cascade of projection, association, and commissural fibers as a whole. With regard to the principles of organization of the neuronal network, the partitioning of remote effects into direct and indirect components as implied by the model may be relevant to the understanding of the functional recovery of noninfarcted brain regions: one should expect that, owing to the comparatively greater autonomy of the more distant

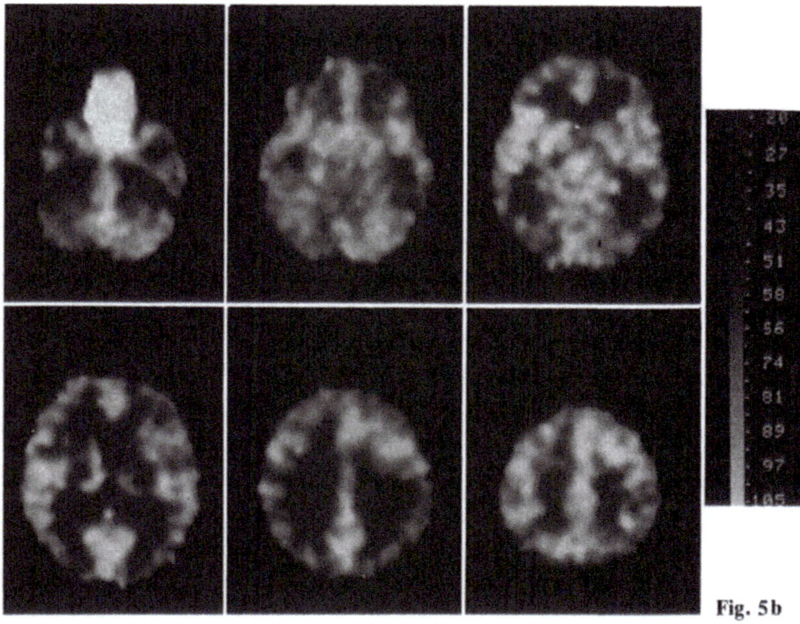

Fig. 5b

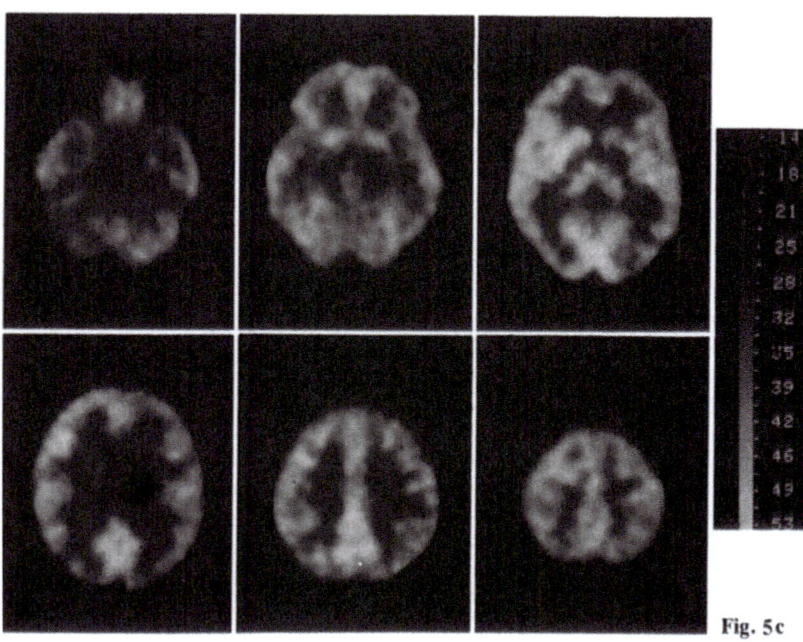

Fig. 5c

Table 4. Multiple regressions of rCMRGl in various morphologically intact brain regions on rCMRGl in the infarcted region and in the contralateral cerebral hemisphere, grouped by lesion site

Localization of lesion	Noninfarcted regions / Coefficients	Ipsilateral: Prefrontal cortex	Midfrontal cortex	Inferior frontal cortex	Insular cortex	Temporopolar cortex	Posterior temporal cortex	Temporoparietal cortex	Parietal cortex	Visual and medial cortex	Occipitopolar cortex	Occipitotemporal cortex	Striatum	Thalamus	Superior brainstem	Contralateral Cerebellum
Midfrontal	r^2	.991	.966	.953	.885	.999	.973	.969	.965	.988	.991	.994	.957	.966	.978	.994
(n = 9)	Std lesion MR	.13	.33	.17	.17	.10*	.16	.35	.37	.03	.07	.26*	.42	.09	.00	.91**
	Std cl hemi MR	.77***	.51**	.59**	.36	1.01***	.65**	.55**	.45*	.79***	1.07***	1.01***	.44*	.74**	.88***	1.30***
Inferior frontal	r^2	.989	.986	.991	.990	.988	.993	.989	.992	.988	.998	.996	.972	.977	.985	.987
(n = 6)	Std lesion MR	.41	.51	.53	.77	.44	.64	.74	.50	−.16	−.49	.10	.64	.55	−.08	.21
	Std cl hemi MR	.30	.21	.37	.06	.34	.24	.10	.32	.64*	1.07***	.78**	.04	.13	.59	.36
Temporoparietal	r^2	.988	.990	.972	.985	.982	.989	.981	.986	.994	.994	.994	.973	.959	.995	.985
(n = 10)	Std lesion MR	.11	.01	−.34	−.19	.51	.36*	.43	.33	−.11	−.01	.04	.15	.53*	.07	.08
	Std cl hemi MR	.68***	.78***	.73***	.97***	.98**	.55***	.62**	.65***	.92***	.85***	.92***	.54**	.15	.90***	.82***
Striatum	r^2	.998	.998	.995	.958	.997	.997	.995	.995	.993	.983	.996	.994	.986	.988	.994
(n = 6)	Std lesion MR	.24	.35	.27	.22	.71**	.17	.20	.26	.21	.11	.23	.40	.17	.02	.52
	Std cl hemi MR	.88***	.73***	.77**	.45	.42**	.71***	.70**	.58**	.58**	.57*	.62**	.56**	1.08	.74**	.59**
Thalamus	r^2	.997	.997	.994	.993	.996	.995	.998	.996	.990	.979	.995	.989	.922	.995	.992
(n = 9)	Std lesion MR	.38*	.14	.38	−.22	.00	.02	.33	.47*	.49	.25	.18	.26	.53	.88**	.72*
	Std cl hemi MR	.68***	.86***	.58***	1.24***	1.11***	1.26	1.36***	.79***	.62***	.74**	1.43***	.76***	.10	.42***	.47**

$^*P \leq 0.05$; $^{**}P \leq 0.01$; $^{***}P \leq 0.001$; r^2, multivariate coefficient of determination; Std lesion MR, standardized regression coefficient of lesion metabolic rate; Std cl hemi MR, standardized regression coefficient of metabolic rate of contralateral ("unaffected") cerebral hemisphere (regression coefficients are standardized by dividing a parameter estimate by the ratio of the sample SD of the dependent variable to the sample SD of the regressor variable) elements of a series of neurons, indirect effects are more likely to recede with time than those effects that depend more directly on the evolution of the ischemic lesion. Accordingly, Kushner et al. [20] found asymmetrical cerebellar hypometabolism resulting from a parietal lesion to resolve with time, while Martin and Raichle [24] described persistent cerebellar asymmetry in patients with frontal lobe infarction.

Metabolism/blood Flow Relationships

In the 14 patients who had a CBF measurement in addition to their FDG study, blood flow and glucose metabolic rates in the cerebral hemisphere contralateral to the infarct averaged 53 ± 16.5 ml/100 g/min and 29.2 ± 5.62 μmol/100 g/min, respectively. Overall, the correspondence between regional blood flow and metabolic rates was poor (Table 5). In view of the heterogeneous time frame — from onset of symptoms to PET study — of the present sample, this finding is not unexpected, considering the well-known variability of flow-to-metabolism coupling during the course of ischemic cerebrovascular disease [3, 19, 22, 41]. After standardization of the two variables with respect to the individual value in a large reference region (the "unaffected" cerebral hemisphere), however, significant correlations were obtained for the majority of noninfarcted regions — even in the injured hemisphere and in the cerebellum contralateral to the ischemic lesion (Table 5). This observation suggests considerable similarity in the topographical expression and relative magnitude of remote effects on rCBF and rCMRGl, despite substantial intersubject variation in that fundamental pair of physiological parameters. Another interesting relationship emerged when the individual ratios of ipsilateral cerebellar to contralateral cerebral hemispheric blood flow and metabolism were compared: the cerebellar-to-cerebral flow ratio (1.11 ± 0.152) was consistently higher ($P < 0.001$, by paired t-test) than the corresponding metabolic ratio. This observation is in excellent agreement with experimental results on autoregulation in various parts of the CNS, suggesting a relatively greater flow reserve of the cerebellum [35]. Together, they may explain why cerebellar infarction is much less common than supratentorial ischemic stroke.

Table 5. Correlations of rCMRGl and rCBF in various noninfarcted brain regions of 14 stroke patients

Region	Spearman correlation coefficients for metabolic and blood flow rates			
	Raw data		Data standardized with respect to "unaffected" cerebral hemisphere	
"Unaffected" cerebral hemisphere	.018	ns		
"unaffected" cerebellar hemisphere	.162	ns	.697	**
Ipsilateral prefrontal cortex	−.222	ns	.578	
Ipsilateral midfrontal cortex	−.237	ns	.574	*
Ipsilateral inferior frontal cortex	.128	ns	.670	**
Ipsilateral insular cortex	.147	ns	.332	ns
Ipsilateral temporopolar cortex	−.035	ns	.512	($P = 0.06$)
Ipsilateral posterior temporal cortex	.302	ns	.921	***
Ipsilateral temporoparietal cortex	.146	ns	.517	($P = 0.06$)
Ipsilateral parietal cortex	−.378	ns	.820	***
Ipsilateral visual and medial cortex	.068	ns	.424	ns
Ipsilateral occipitopolar cortex	.403	ns	.440	ns
Ipsilateral occipitotemporal cortex	.079	ns	.626	*
Ipsilateral striatum	.084	ns	.798	***
Ipsilateral thalamus	.165	ns	.899	***
Ipsilateral superior brainstem	.446	ns	.574	*
Contralateral cerebellum	−.121	ns	.635	*

ns, not significant; *, $P < 0.05$; **, $P < 0.01$; ***, $P < 0.001$

Conclusions

Degree and topographical distribution of functional depression of CNS areas remote from the site of an ischemic lesion seem to depend mainly on three factors: infarct location, residual lesion metabolic rate, and time. Diaschistic effects may be partitioned into direct and indirect components as described by a simple, yet highly significant, linear model. Direct effects, in particular, are characteristic of only a few brain regions. While all remote effects manifest themselves predominantly in the contralateral cerebellar hemisphere and on the ipsilateral side of the rest of the brain, both metabolically and hemodynamically, global effects on the brain as a whole are common. Specific regulatory flow mechanisms seem to afford the cerebellum extensive protection against ischemic damage.

References

1. Baron JC, Bousser MG, Comar D, Castaigne P (1980) "Crossed cerebellar diaschisis" in human supratentorial brain infarction. Ann Neurol 8:128
2. Baron JC, Bousser MG, Comar D, Duquesnoy N, Sastre J, Castaigne P (1981) "Crossed cerebellar diaschisis": A remote functional depression secondary to supratentorial infarction of man. J Cereb Blood Flow Metab 1 [Suppl]: 500–501
3. Baron JC, Rougemont D, Soussaline F, Bustany P, Crouzel C, Bousser MG, Comar D (1984) Local interrelationships of cerebral oxygen consumption and glucose utilization in normal subjects and in ischemic stroke patients: a positron tomography study. J Cereb Blood Flow Metab 4:140–149
4. Bergström M, Litton J, Eriksson L, Bohm C, Blomqvist G (1982) Determination of object contour from projections for attenuation correction in cranial positron emission tomography. J Comput Assist Tomogr 6:365–372
5. Bergström M, Eriksson L, Bohm C, Blomqvist G, Litton J (1983) Correction for scattered radiation in a ring detector positron camera by integral transformation of the projections. J Comput Assist Tomogr 7:42–50
6. Celesia GG, Polcyn RE, Holden JE, Nickles RJ, Koeppe RA, Gatley SJ (1984) Determination of regional cerebral blood flow in patients with cerebral infarction — use of fluoromethane labeled with fluorine 18 and positron emission tomography. Arch Neurol 41:262–267
7. Duara R, Grady C, Haxby J, Ingvar D, Sokoloff L, Margolin RA, Manning RG, Cutler NR, Rapoport SI (1984) Human brain glucose utilization and cognitive function in relation to age. Ann Neurol 16:702–713
8. Eriksson L, Bohm C, Kesselberg M, Blomqvist G, Litton J, Widen L, Bergström M, Ericson K, Greitz T (1982) A four ring positron camera system for emission tomography of the brain. IEEE Trans Nucl Sci 29:539–543
9. Frackowiak RSJ, Lenzi GL, Jones T, Heather JD (1980) The quantitative measurement of regional cerebral blood flow and oxygen metabolism in man using oxygen-15

and positron emission tomography: theory, procedure and normal values. J Comput Assist Tomogr 4:727–736
10. Heiss WD, Vyska K, Kloster G, Traupe H, Freundlieb C, Hoeck A, Feinendegen LE, Stoecklin G (1982) Demonstration of decreased functional activity of visual cortex by (^{11}C)methylglucose and positron emission tomography. Neuroradiology 23:45–47
11. Heiss WD, Ilsen HW, Wagner R, Pawlik G, Wienhard K (1983) Remote functional depression of glucose metabolism in stroke and its alteration by activating drugs. In: Heiss WD, Phelps ME (eds) Positron emission tomography of the brain. Springer, Berlin Heidelberg New York, pp 162–168
12. Heiss WD, Pawlik G, Herholz K, Wagner R, Göldner H, Wienhard K (1984) Regional kinetic constants and cerebral metabolic rate for glucose in normal human volunteers determined by dynamic positron emission tomography of (^{18}F)-2-fluoro-2-deoxy-D-glucose. J Cereb Blood Flow Metab 4:212–223
13. Herholz K, Pawlik G, Wienhard K, Heiss WD (1985) Computer assisted mapping in quantitative analysis of cerebral positron emission tomograms. J Comput Assist Tomogr 9:154–161
14. Hoedt-Rasmussen K, Skinhoj E (1964) Transneural depression of the cerebral hemispheric metabolism in man. Acta Neurol Scand 40:41–46
15. Holden JE, Gately SJ, Nickles RJ, Koeppe RA, Celesia GG, Polcyn RE (1983) Regional cerebral blood flow measurement with fluoromethane and positron tomography. In Heiss WD, Phelps ME (eds) Positron emission tomography of the brain. Springer, Berlin Heidelberg New York, pp 90–94
16. Jones T, Chesler DA, Ter-Pogossian MM (1976) The continuous inhalation of oxygen-15 for assessing regional oxygen extraction in the brain of man. Br J Radiol 49:339–343
17. Kempinsky WH (1958) Experimental study of distant effects of acute focal brain injury. Arch Neurol Psychiatr 79:376–389
18. Koeppe RA, Holden JE, Polcyn RE, Nickles RJ, Hutchins GD, Weese JL (1985) Quantitation of local cerebral blood flow and partition coefficient without arterial sampling: theory and validation. J Cereb Blood Flow Metab 5:214–223
19. Kuhl ED, Phelps ME, Kowell AP, Metter EJ, Selin C, Winter J (1980) Effects of stroke on local cerebral metabolism and perfusion: mapping by emission computed tomography of ^{18}FDG and $^{13}NH_3$. Ann Neurol 8:47–60
20. Kushner M, Alavi A, Reivich M, Dann R, Burke A, Robinson G (1984) Contralateral cerebellar hypometabolism following cerebral insult: a positron emission tomographic study. Ann Neurol 15:425–434
21. Lenzi GL, Frackowiak RSJ, Jones T (1981) Regional cerebral blood flow (CBF) oxygen utilisation ($CMRO_2$) and oxygen extraction ratio (OER) in acute hemispheric stroke. J Cereb Blood Flow Metab 1 [Suppl 1]:504–505
22. Lenzi GL, Frackowiak RSJ, Jones T (1982) Cerebral oxygen metabolism and blood flow in human cerebral ischemic infarction. J Cereb Blood Flow Metab 2:321–335
23. Lilliefors HW (1967) On the Kolmogorov-Smirnov test for normality with mean and variance unknown. J Am Statist Assoc 62:399–402
24. Martin WRW, Raichle ME (1983) Cerebellar blood flow and metabolism in cerebral hemisphere infarction. Ann Neurol 14:168–176
25. Melamed E, Lavy S, Portnoy Z (1975) Regional cerebral blood flow response to hypocapnia in the contralateral hemisphere of patients with acute cerebral infarction. Stroke 6:503–508
26. Metter EJ, Riege WH, Kuhl DE, Phelps ME (1984) Cerebral metabolic relationships for selected brain regions in healthy adults. J Cereb Blood Flow Metab 4:1–7

27. Meyer JS, Shinohara Y, Kanda T, Fukuuchi Y, Ericsson AD, Kok NK (1970) Diaschisis resulting from acute unilateral cerebral infarction. Arch Neurol 23:241–247
28. Paschen W, Sato M, Pawlik G, Umbach C, Heiss WD (1985) Neurologic deficit, blood flow, and biochemical sequelae of reversible focal cerebral ischemia in cats. J Neurol Sci 68:119–134
29. Patronas NJ, DiChiro G, Smith BH, DeLaPaz R, Brooks RA, Milam HL, Kornblith PL, Bairamian D, Mansi L (1984) Depressed cerebellar glucose metabolism in supratentorial tumors. Brain Res 291:93–101
30. Phelps ME, Hoffman EJ, Mullani NA, Ter-Pogossian MM (1975) Application of annihilation coincidence detection to transaxial reconstruction tomography. J Nucl Med 16:210–223
31. Phelps ME, Huang SC, Hoffman EJ, Selin CJ, Sokoloff L, Kuhl DE (1979) Tomographic measurement of local cerebral clucose metabolic rate in humans with (F-18)2-fluoro-2-deoxy-D-glucose: validation of method. Ann Neurol 6:371–388
32. Phelps ME, Mazziotta JC, Kuhl DE, Nuwer M, Packwood J, Metter J, Engel Jr J (1981) Tomographic mapping of human cerebral metabolism: visual stimulation and deprivation. Neurology (Minneap) 31:517–529
33. Reivich M, Kuhl D, Wolf A, Greenberg J, Phelps ME, Ido T, Casella V, Fowler J, Hoffman E, Alavi A, Som P, Sokoloff L (1979) The (^{18}F)-fluorodeoxyglucose method for the measurement of local cerebral glucose utilization in man. Circ Res 44:127–137
34. Sato M, Paschen W, Pawlik G, Heiss WD (1984) Neurologic deficit and cerebral ATP depletion after temporary focal ischemia in cats. J Cereb Blood Flow Metab 4:173–177
35. Sato M, Pawlik G, Heiss WD (1984) Comparative studies of regional CNS blood flow autoregulation and responses to CO_2 in the cat — effects of altering arterial blood pressure and $PaCO_2$ on rCBF of cerebrum, cerebellum, and spinal cord. Stroke 15: 91–97
36. Slater R, Reivich M, Goldberg H, Banka R, Greenberg J (1977) Diaschisis with cerebral infarction. Stroke 8:684–690
37. Ter-Pogossian MM, Phelps ME, Hoffman EJ, Mullani NA (1975) A positron emission transaxial tomograph for nuclear medicine imaging (PETT). Radiology 14:89–98
38. von Monakow C (1914) Die Lokalisation im Großhirn und der Abbau der Funktion durch kortikale Herde. Bergmann, Wiesbaden
39. Wagner R (1984) A fast, high yield synthesis of ^{18}F-fluoromethane from $^{18}F\text{-}F_2$. J Label Comp Radiopharm 21:1229–1230
40. Wienhard K, Pawlik G, Herholz K, Wagner R, Heiss WD (1985) Estimation of local cerebral glucose utilization by positron emission tomography of (^{18}F)2-fluoro-2-deoxy-D-glucose: a critical appraisal of optimization procedures. J Cereb Blood Flow Metab 5:115–125
41. Wise RJS, Bernardi S, Frackowiak RSJ, Legg NJ, Jones T (1983) Serial observations on the pathophysiology of acute stroke — the transition from ischaemia to infarction as reflected in regional oxygen extraction. Brain 106:197–222

Focal Flow Disturbances in Acute Strokes: Effects on Regional Metabolism and Tissue pH

Y. L. YAMAMOTO, A. M. HAKIM, M. DIKSIC, R. P. POKRUPA, E. MEYER, J. TYLER, A. C. EVANS, K. WORSLEY, C. J. THOMPSON and W. H. FEINDEL

Introduction

In acute stroke, the reduction of blood flow causes tissue damage and sets of a complex escalating chain reaction from acid metabolites and also from products of disturbed intracellular ionic homeostasis. These events may lead to lipolysis and proteolysis [35, 41]. Although regional cerebral blood flow (rCBF) is an appropriate function to measure in acute stroke, recent studies have revealed that it varies widely in patients with similar neurologic deficits from stroke. Moreover, ^{133}Xe intracarotid studies have shown focal hyperemia to be a striking early feature in some patients with stroke [29, 42–44] and for this reason also, rCBF is a poor indicator of residual cerebral tissue function. In contrast, positron emission tomography (PET) imaging has shown a consistently reduced regional cerebral metabolic rate (below 67 μmol/100 g/min) for oxygen ($rCMRO_2$) in the infarcted area [30, 48, 49].

Focal hyperemia associated with low $rCMRO_2$ and low oxygen extraction fraction (OEF) results in hyperoxic venous blood. This phenomenon of "red veins," first reported by Penfield et al. [32] from observations during brain operations of the circulation in the postictal state, was later described in relation to tumors and longstanding infarcts by Feindel and Perot [12]. The latter further established that the appearance of hyperoxic venous blood was due to (a) structural preferential shunting through collaterals, or (b) "metabolic" shunting, where tissue uptake of oxygen was reduced because of decreased metabolic demand of damaged brain tissue. The most appropriate term to describe this increase in blood flow is "reactive hyperemia." Lassen [29] coined the term "luxury perfusion" for this increased flow associated with early cerebral infarction, but physiologically the increased flow is often related more to a poverty of perfusion.

PET has greatly increased our understanding of the pathophysiology of stroke by making it possible to measure several mismatching patterns between rCBF and metabolic parameters [1–3, 31]. Recently, the use of ^{11}C-labelled DMO (5,5-dimethyl-2,4-oxazolidinedione) to measure brain pH has added further information on acute stroke [19, 47].

We measured rCBF, rOEF, $rCMRO_2$, regional cerebral metabolic rate for glucose (rCMRGl), and brain pH in patients within 48 h of the clinical onset of stroke [17–19]. We now report on relative focal hyperemia and its relationship with other metabolic features in cerebral infarct as observed in 12 patients.

Methods

Patients

We carried out PET studies on 12 adult patients within 48 h of the clinical onset of acute stroke in the territory of the internal carotid artery. CT scans revealed there were no hemorrhagic or nonischemic lesions. All patients had lesions in the cerebral hemisphere with focal neurologic deficits that persisted throughout the studies. Clinical data are summarized in Table 1. Of the 12 patients, 4 were diabetic; however, case 10 had only mild diabetes with 179 mg/100 ml plasma glucose during the PET studies. Four patients had atrial fibrillation.

PET Procedures

PET images were obtained on the Therascan-3128, a two-ring tomographic positron emission scanner. Each ring contained 64 bismuth germanate detectors and defined three image planes. The resolution of the scanner, full width at half maximum, was 12 mm [6]. The slice thickness (axial resolution) was 11–13 mm.

The quantitative recovery of the system was evaluated using the double isotope method [6, 7] and contrast recovery phantoms [46]. An explicit deconvolution procedure removed scattered events from the raw data before reconstruction [7, 46]. The patient's head was held fixed by a thermoplastic mask. A reference position was established by

Table 1. Patient Data

Group no.	Case no.	Age/Sex (yr)	Stroke to PET studies (h)	BP (mmHg)	Plasma glucose (mg/100 ml)	$PaCO_2$ (torr)	CT findings within 24 h	CT findings 1 week later	Other medical history
I	1	61/F	24–31	130/80	115	39	Normal	L. Parietal infarct	—
	2	60/M	23–44	170/70	104	40	R. ACA infarct		Cardiac irregularity
	3	59/M	24–36	170/90	132	35	Normal	L. Parietal infarct	RIND
	4	57/F	24–48	140/80	289–319	41	Normal	L. MCA infarct	Diabetes Atrial fibrillation
II	5	73/F	16–30	160/80	227	35	Normal	L. MCA + ACA + PCA infarct	Diabetes, chronic heart disease
	6	71/F	26–35	190/85	120	41	Normal	L. Hemisph. infarct	Hypertension
	7	75/F	31–45	140/80	284–295	39	Normal	L. MSA + ACA infarct	Diabetes, Atrial fibrillation
	8	71/F	23–28	160/100	133	34	Normal	L. MCA + ACA infarct	Atrial fibrillation
III	9	57/M	18–42	180/100	70	38	Small R. parietal		Hypertension
	10	86/F	24–30	170/90	179	45	Normal	Incomplete L. MCA infarct	Diabetes, Atrial fibrillation
	11	28/M	16–22	160/100	102	35	Normal	Normal	—
	12	74/M	23–48	140/80	145	36	Normal	L. MCA infarct	Abdominal aortic aneurysm

–Hyperglycemia (>200 mg/100 ml); L., left; R., right; MCA, middle cerebral artery; ACA, anterior cerebral artery; PCA, posterior cerebral artery: RIND, reversible ischemic neurologic deficit; BP, blood pressure aligning a wall-mounted laser with the patient's orbitomeatal line. PET images were obtained 30, 42, 50, 54, 62, and 74 mm above the orbitomeatal line in all studies. During the scan the head position was monitored and movement restricted to ensure that positions would be reproducible and comparable from one study to another.

After reconstruction into a 128 × 128 image matrix using the filtered back-projection method, static images were converted from distribution of radioactivity to functional maps by correlating the radioactivity maps with blood and plasma concentration measurements. Numerical print-outs of the data in a compacted 64 × 64 matrix format were also obtained.

^{15}O Studies for rCBF, rOER, $rCMRO_2$, and Regional Cerebral Blood Volume (rCBV)

Oxygen-15-labelled gases, ^{15}O-O_2, ^{15}O-CO_2, and ^{15}O-CO, were administered sequentially in one session to provide three separate data sets, which were incorporated to produce functional maps of rCBF, rOEF or regional oxygen extraction ratio (rOER), $rCMRO_2$, and regional cerebral blood volume (rCBV), following the equations of Frackowiak et al. [14]. During all studies arterial samples were repeatedly collected for measurement of blood gases, pH, hematocrit, and radioactivity concentration.

1. ^{15}O-O_2 study. Molecular ^{15}O was produced by deuteron irradiation of N_2 gas with 4% O_2 carrier in a continuous flow target arrangement. Impurities were removed by passing the gas through a soda lime trap and an activated charcoal trap. Radiochemical purity was 99%. The gas was mixed with medical air for continuous administration through a light disposable plastic mask at a rate of 10 mCi/min and 80 ml/min. Once inhalation began, a 12-min period was allowed for the establishment of equilibrium before two 10-min scans were performed at the fixed levels. Typically, 1.5–2.5 million true events per image were recorded.

2. ^{15}O-CO_2 studies. Oxygen-15-labelled carbon dioxide was produced by deuteron irradiation of 2.5% CO_2 in N_2 gas. Impurities were removed by passing the gas through an activated charcoal trap. The administration and scanning procedures were similar to those used in the ^{15}O-O study. The delivery rate was 5 mCi/min at a flow of 60 ml/min, and typically 2–4 million true events per image were recorded.

3. ^{15}O-CO study. Oxygen-15-labelled CO was produced by passing ^{15}O-labelled molecular oxygen, produced as described above, through a column of activated charcoal at 900°C. Impurities were removed by means of a soda lime column. Irradiated gas containing less than 1% of ^{15}O-CO was delivered to the PET room at 80 mCi/min and a flow of 100 ml/min, and mixed with medical air delivered at a rate of 200 ml/min. The mixture was delivered for about 4 min, followed by a 1-min equilibration delay before two low-resolution, 1-min scans were collected. Typical images contained 300000–500000 true events.

The oxygen analysis package generated a map for each of the four functional parameters of rCBF, rOEF, rCBV, and $rCMRO_2$ for each anatomical plane as composite dataset.

rCMRGl Study Using [^{18}F]-Fluorodeoxyglucose (FDG)

^{18}F-FDG was prepared by reacting in situ ^{18}F-acetylhypofluorite with triacetyl-D-glucal. After extraction with CH_2Cl_2 and hydrolysis with 2 *N*-HCl, ^{18}F-FDG was dissolved in a 1 *N* phosphate buffer (pH = 7). The specific activity after 30 min of irradiation and a 60-min synthesis was approximately 685 mCi/mmol [10]. Recently we observed, as reported by Bida et al., [4] (unpublished data), that the final radiopharmaceutical contains on average 15% of 2-deoxy-2-fluoro-D-mannose. The influence of this impurity on glucose utilization was assessed by a simulation using the rate constants for 2-deoxy-2-fluoro-D-mannose measured in normal volunteers [38] and the same impurity level. The simulation showed that this impurity would influence rCMRGl less than 3% (unpublished data).

A calculation of rCMRGl was made using the Phelps et al. [34] adaptation to human studies of the Sokoloff et al. [45] method, incorporated with a reformulation of the Phelps operational equation suggested by Brooks [5]. A 5-mCi bolus of ^{18}F-FDG was injected into the antecubital vein, and blood samples were taken from the opposite arm, using either direct arterial sampling or samples from a vein arterialized by immersing the hand in a water-filled glove box at 42°C [34]. The sampling schedule yielded 10–15 1-ml samples over the first 2 min, with gradually extended intervals for the remainder of the study for a total of 25–30 samples. During the uptake phase, ambient noise and light levels were kept to a minimum, but no eye blinds or ear plugs were used. Once equilibrium was established at about 45 min, two 10-min static scans were performed to yield 6 slices over the ischemic levels. Typical images contained 3–6 million true events. After reconstruction, static images were converted from distributions of radioactivity to functional maps by correlation of the radioactivity images with plasma concentration measurements. Values for all rate constants and for the lumped constant were those established by Huang et al. [25]. They were not modified to account for the changes in the rate constants that occur with ischemia [20, 22].

Cerebral pH Measurement with ^{11}C-DMO

The synthesis of ^{11}C-5,5-dimethyl-2,4-oxazolidinedione has been described in detail elsewhere [9]. To a solution of about 50 mg of 2-hydroxy-2-methylpropionamide, prepared by a published procedure [15] in 3 ml of dry acetonitrile, about 50 mg of powdered KOH was added. The suspension was cooled to −5°C and ^{11}C-labelled phosgene introduced for about 20 min [11]. At the end of phosgene introduction the solvent was evaporated, the residue dissolved in 1 ml of water, and the pH adjusted to 8. The solution was passed through a Sep-Pak C_{18}-column and eluted with 10 ml of a methanol-water (3:2) mixture. After evaporation of the elution solvent to about 0.5 ml, ^{11}C-labelled DMO was dissolved in a buffer (pH = 7), and sterilized through a 0.2 μm millipore filter. It had been noted that exclusion of moisture was essential if material of high radiochemical purity was to be obtained. The presence of moisture prevented the closure of the oxazolidine ring, giving a compound with phosgene bound to the amino group of 2-hydroxy-2-methylpropionamide. This contaminant was found to remain unchanged in the cerebrovascular compartment for the duration of this study (unpublished data).

The cerebral pH was determined using the method of Kearfott et al. [26] and Rottenberg et al. [39]. Carbon-11-labelled DMO was injected as a 20-mCi bolus. DMO, a weak acid, and nonionized DMO (HDMO), were assumed to be freely diffusible between the vascular, extracellular, and intracellular spaces. At equilibrium the concentration of HDMO was assumed to be constant everywhere. Thus, the Henderson-Hasselbalch equation can be used with the observed tissue and plasma concentrations and the plasma pH to yield tissue pH. While waiting 45 min for the DMO to equilibrate, a dynamic data collection was initiated at injection with the following scan sequence: 6 scans every 30 s, then 11 scans every minute, then 3 scans every 2 min, and finally 4 scans every 5 min, to examine equilibration of ^{11}C-DMO in the ischemic tissue. At equilibrium, two three slice static scans of 10-min duration were performed. Blood samples were taken during each scan to obtain an average value for plasma pH and plasma ^{11}C-radioactivity. After reconstruction, pH maps were produced according to the equation:

$$\mathrm{rCpH} = \log_{10}(\mathrm{R}(10^{\mathrm{pHb}} + 10^{\mathrm{pKa}}) - 10^{\mathrm{pKa}})$$

where rCpH represents tissue pH, pHb represents plasma pH, pKa represents pK for DMO, and R represents ratio of tissue to plasma activity concentrations.

Data Analysis

For all analyses, a 64 × 64 matrix numerical printout was used, in which each number represented the mean of a 2 × 2 pixel area (one pixel is 2 mm × 2 mm). The rCBF printouts from one to two levels in each patient were analyzed numerically to define the region of interest (ROI) corresponding to the location of the cerebral infarct identified in the second CT scan findings (Table 1). This was done on the basis of rCBF focal abnormalities, which were correlated with the CT scan finding. The appropriate ROI was defined and outlined by the overlaying of a transparent plastic sheet. The mirror image ROI was also defined in the contralateral homologous tissue. The appropriate ROIs were then transcribed onto each of the functional PET printouts, and the corresponding mean values of rCBF, $\mathrm{rCMRO_2}$, rCMRGl, rOEF, rCBV, and rCpH were determined.

The ROIs were 1.6–3.2 cm^2 in area and contained not only gray matter, but also a certain proportion of white matter.

Areas with abnormally high or low rCBF were determined in the following way. Values of rCBF in the nonaffected contralateral homologous areas in all 11 patients were first determined together with their standard deviation (SD). Then ROI areas (1.6–3.2 cm^2) with rCBF values 2 SD or more above the mean of contralateral rCBF values were defined as focal hyperemic areas. Areas with rCBF values of 2 SD or more below the mean contralateral rCBF value were defined as ischemic areas.

Thus, focal hyperemia and ischemia in this study were defined relative to the mean rCBF values in the nonaffected contralateral homologous areas.

Since the replications are obviously positively correlated, their SD will underestimate that of the observations. Instead, for a particular variable, it was assumed that the control sites had the same true mean across patients, and that the disease and control sites were independent. Thus, the SD of the disease/control differences can be estimated by $\sqrt{2}$ times the SD of the control sites. Even if the disease and control sites are positively correlated, as is likely to be the case, this estimate will be conservative.

To test for a disease/control difference, the *t* statistic is the difference divided by the estimated SD, with 7 degrees of freedom (6 for CpH).

Results

The initial investigation of rCBF changes in the lesion as compared to the contralateral cerebral hemisphere in all 12 patients revealed that 4 (36%) of 11 patients had focal hyperemia, and 7 (64%) of 11 patients had focal ischemia. With case 12, technical difficulty owing to head movement and irregular respiratory patterns during the ^{15}O studies resulted in unreliable data, which were not included in this study. On the basis of the argument of Lenzi et al. [30], and as confirmed by follow-up CT scans in our study [19], that an $rCMRO_2$ value of less than 67 μmol/100 g/min will lead to infarction, three subgroups were distinguished in this study. Group I consists of 4 patients who had significantly increased rCBF in the lesions as compared to the contralateral homologous region, and $rCMRO_2$ values below 67 μmol/100 g/min, as shown in Table 2. Group II consists of 4 patients who had significantly decreased rCBF in

Table 2. Group I (focal hyperemia)

Metabolic parameters		Case no. (mean ± SD)				Group mean ± SD
		1 $n = 15$	2 $n = 10$	3 $n = 15$	4Δ $n = 15$	
rCBF	D	42 ± 3	31 ± 2	40 ± 2	42 ± 4*	39 ± 5 (+69%)*
(ml/100 g/min)	C	25 ± 2	24 ± 2	24 ± 2	20 ± 1	23 ± 2
$rCMRO_2$	D	35 ± 20*	31 ± 13*	65 ± 10	45 ± 7	44 ± 15 (−56%)*
(μmol/100 g/min)	C	111 ± 10	101 ± 6	105 ± 6	82 ± 5	100 ± 12
rOEF	D	11 ± 6*	20 ± 9*	17 ± 1*	10 ± 2*	14 ± 5 (−69%)*
(%)	C	49 ± 7	45 ± 2	44 ± 3	43 ± 3	45 ± 3
rCBV	D	4.4 ± 0.5	3.7 ± 0.6	3.9 ± 0.8	4.1 ± 0.7	4.0 ± 0.3 (+5%)
(ml/100 g)	C	4.0 ± 0.8	2.9 ± 0.2	3.3 ± 0.7	4.9 ± 0.8	3.8 ± 0.8
rCMRGl	D	33 ± 3	20 ± 2	25 ± 1	20 ± 2	24 ± 6 (−4%)
(μmol/100 g/min)	C	34 ± 5	25 ± 2	26 ± 2	15 ± 2	25 ± 8
rCpH	D	7.01 ± 0.05	–	7.02 ± 0.06	7.05 ± 0.06	7.05 ± 0.04 (+1%)
	C	6.94 ± 0.07	–	6.97 ± 0.12	6.99 ± 0.07	6.97 ± 0.02

D, diseased site; C, contralateral homologous site; *, statistical significance *($P < 0.05$); Δ, hyperglycemia (> 200 mg/100 ml)* the lesion as compared to the contralateral homologous region, and less than 67 μmol/100 g/min of $rCMRO_2$, as shown in Table 3. Group III consists of 3 patients who had lower rCBF values in the lesion than in the contralateral homologous regions, and $rCMRO_2$ values higher than 67 μmol/100 g/min, which have not been discussed further in this study.

Group I (Focal Hyperemia)

Four (36%) of 11 patients belonged in this group. The group mean value for rCBF in the lesions was 39 ± 5 ml/100 g/min, as compared to the group mean value for rCBF in the contralateral homologous regions of 23 ± 2 ml/100 g/min, indicating a 69% increase ($P < 0.02$). Values of $rCMRO_2$ in the lesion in all 4 cases were below 65 μmol/100 g/min, indicating the presence of irreversible tissue damage in these areas. The group mean value of $rCMRO_2$ was 44 ± 15 μmol/100 g/min in the infarcts, as compared to 100 ± 12 μmol/100 g/min in the contralateral homologous areas, indicating a 56% reduction ($P < 0.02$).

Values of rOEF were also markedly reduced in the infarcted area. The group mean value was 14 ± 5% as compared to 45 ± 3% in the contralateral area, indicating a 69% reduction ($P < 0.005$). There was no significant change in either rCBV or rCMRGl in the acute stage of these infarcted areas.

The rCpH in the infarcted area in all patients, including case 4, who had hyperglycemia of 289–319 mg/100 ml, was slightly but consistently higher than in the contralateral homologous regions, arguing against the presence of acid metabolites in these infarcted areas. These metabolic changes related to focal hyperemia in the cerebral infarct are illustrated in Fig. 1A–D. Therefore, the cerebral infarcted areas with associated focal hyperemia undergo a significant reduction in rOEF and $rCMRO_2$, associated with a slight increase in rCpH, indicating the presence of hyperoxygenated cerebral venous blood without significant lactoacidosis. This is suggestive of physiological or metabolic shunting [12, 13, 50, 51].

The clinical and laboratory investigations in all patients in this group indicated that case 4 had a history of atrial fibrillation, and case 2 a long history of cardiac arrhythmia. The remaining two patients did not have a history of cardiac disorder to suggest embolic episodes.

Table 3. Group II (focal ischemia)

Metabolic parameters		Case no. (mean ± SD)				Group mean ± SD
		5Δ n = 10	6 n = 15	7Δ b = 20	8 n = 20	
rCBF	D	1 ± 0.8*	13 ± 2	4 ± 2	3 ± 0.7	6 ± 5 (−78%)*
(ml/100 g/min)	C	41 ± 3	25 ± 1	24 ± 2	24 ± 2	28 ± 8
$rCMRO_2$	D	20 ± 10*	41 ± 16*	15 ± 10*	15 ± 3*	23 ± 12 (−82%)*
(μmol/100 g/min)	C	149 ± 7	134 ± 9	118 ± 16	124 ± 6	131 ± 13
rOEF	D	135 ± 26*	74 ± 4	66 ± 4	46 ± 4	80 ± 38 (+63%)
(%)	C	35 ± 2	54 ± 4	51 ± 6	55 ± 3	49 ± 9
rCBV	D	1.2 ± 0.4	3.5 ± 0.6	3.9 ± 0.9	2.8 ± 0.3	2.8 ± 1.2 (−12%)
(ml/100 g)	C	3.3 ± 0.3	3.3 ± 0.2	3.9 ± 0.8	2.2 ± 0.4	3.2 ± 0.7
rCMRGl	D	0	7 ± 4	5 ± 5	0.8 ± 0.8*	3 ± 3 (−87%)*
(μmol/100 g/min)	C	15 ± 0.7	28 ± 1	25 ± 3	28 ± 3	24 ± 6
rCpH	D	6.52 ± 0.19*	6.92 ± 0.07	6.74 ± 0.18*	6.95 ± 0.14	6.78 ± 0.2 (−3%)
	C	7.02 ± 0.04	6.98 ± 0.09	7.11 ± 0.08	6.96 ± 0.07	7.02 ± 0.06

D, diseased site; C, contralateral homologous site; *, statistical significance ($P < 0.05$); Δ, Hyperglycemia (> 200 mg/100 ml)

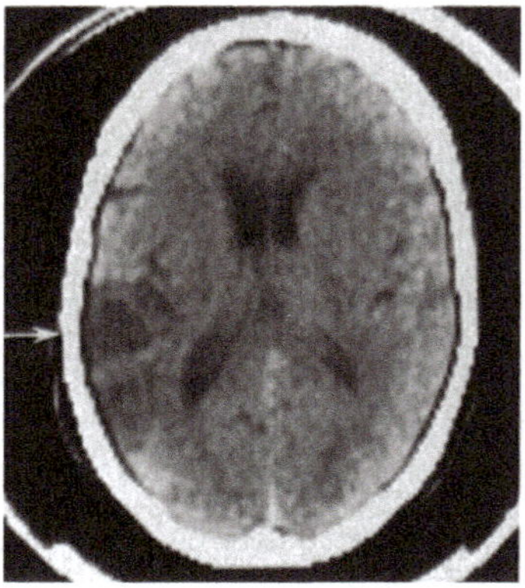

A

Fig. 1 A–D. Case 1. A 61-year-old woman after sudden onset of right hemiplegia and aphasia. **A** CT scan 10 days after the stroke, showing a focal hypodensity in the left parietotemporal region (*arrow*) ; **B** Oxygen-15 studies within 31 h after the stroke, showing a focal increase in rCBF, focal reduction in rOER, slightly increased rCBV, and marked reduction in $rCMRO_2$ in the left parietal region (*arrow*); **C** ^{18}F-FDG study within 31 h after the stroke, showing no significant changes in rCMRGl in the left parietal region (*arrow*); **D** ^{11}C-DMO study within 31 h after the stroke, showing a slightly increased rCpH in the left parietal region (*arrow*)

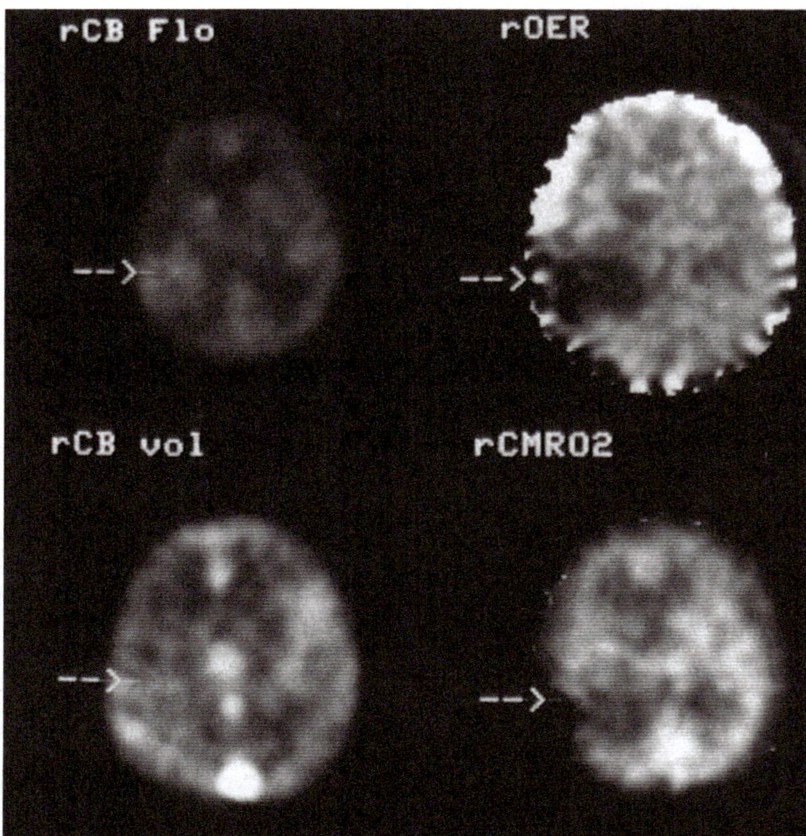

B

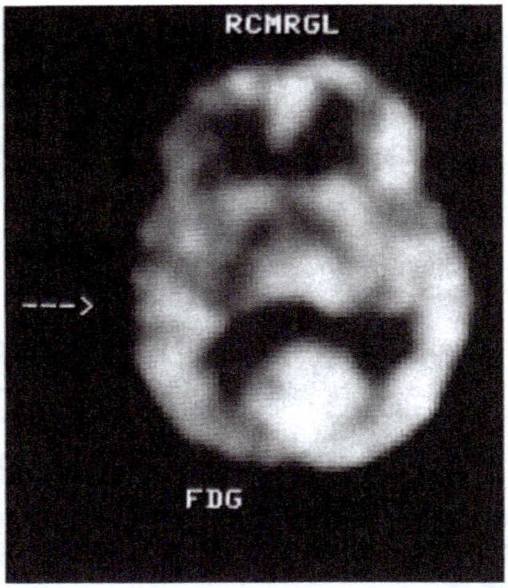

C

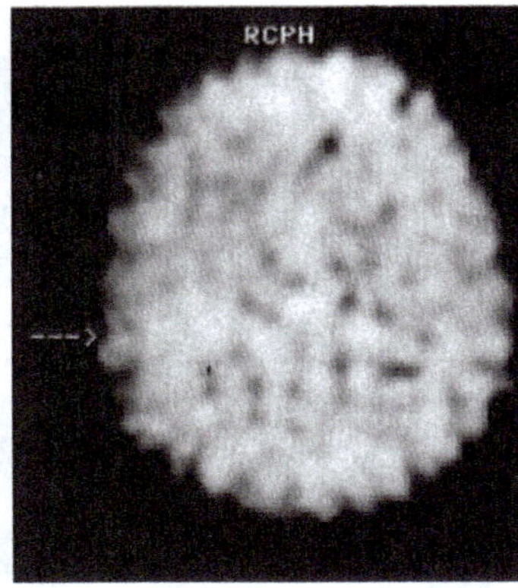

D

Group II (Focal Ischemia)

Four (36%) of 11 patients belonged in this group. The group mean value of rCBF in the infarcted area was 6 ± 5 ml/100 g/min, as compared to the contralateral homologous areas, where the group mean value was 28 ± 8 ml/100 g/min ($P < 0.05$). Values of $CMRO_2$ in the infarcted area were below 41 μmol/100 g/min in all patients, while the group mean value of $rCMRO_2$ was 23 ± 12 μmol/100 g/min, as compared to the contralateral homologous areas, where the group mean value was 131 ± 3 μmol/100 g/min, indicating an 82% reduction ($P < 0.001$). There was a general increase in rOEF in the infarcted areas in three of the four patients. However, there was no significant increase in rOEF in the group mean value in the infarcted areas as compared to the contralateral homologous areas.

The unrealistically high value of rOEF (135%) in the infarcted area in case 5 is caused by the large error inherent in the calculation of rOEF, which essentially consists in the division of the $^{15}O\text{-}O_2$ by the $^{15}O\text{-}CO_2$ image, for low perfusion areas (low-count density).

Values of rCMRGl were significantly reduced in all patients. The group mean value of rCMRGl was only 3 ± 3 μmol/100 g/min in the infarcted areas, as compared to the contralateral homologous areas, where the group mean value was 24 ± 6 μmol/100 g/min, indicating an

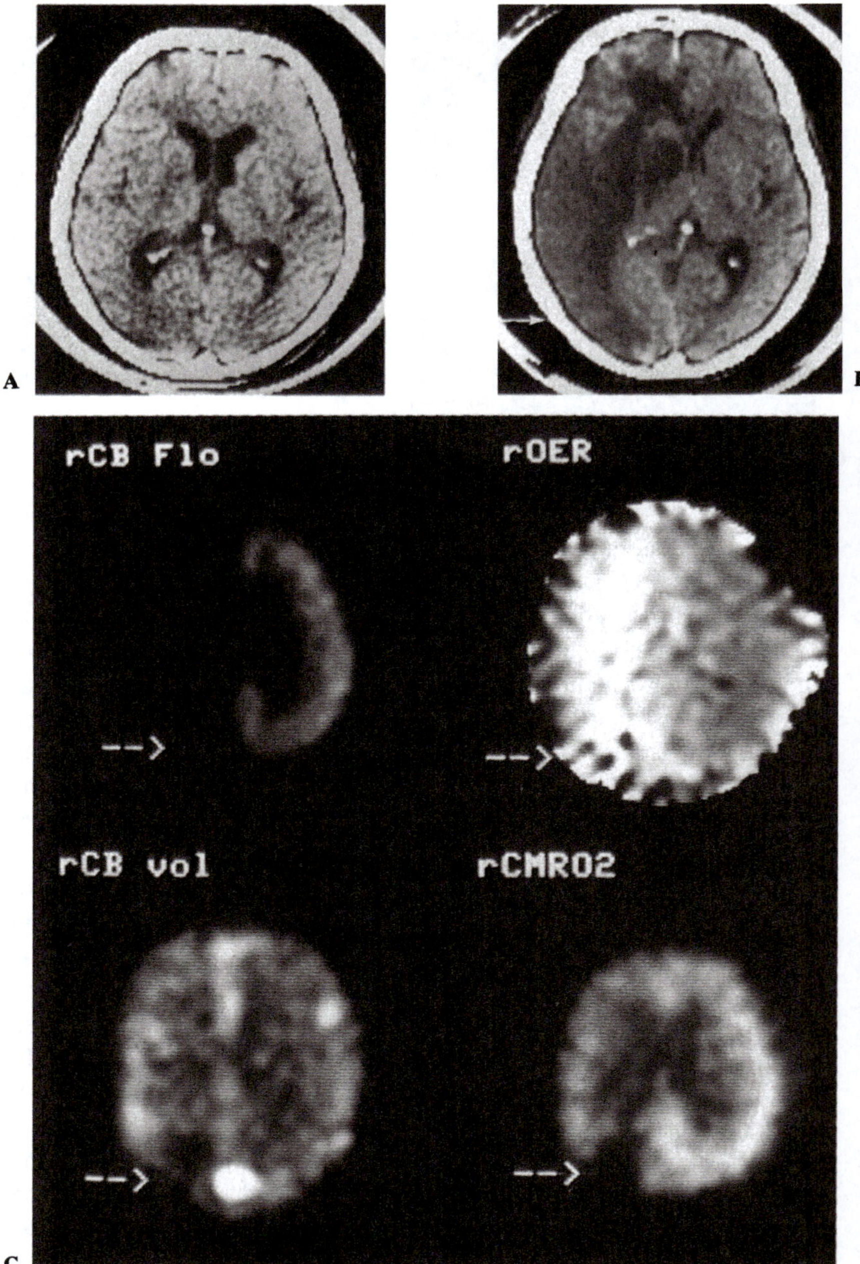
A
B
rCB Flo
rOER
rCB vol
rCMRO2
C

D 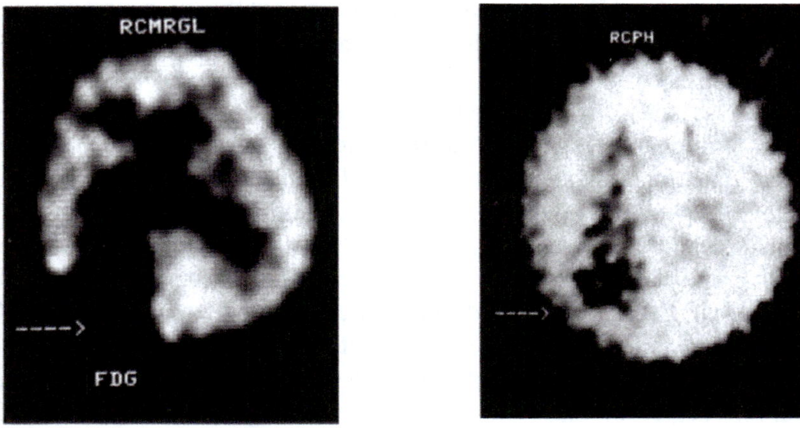 E

Fig. 2A–E. Case 5. A 73-year-old woman after sudden onset of a right hemiplegia, hemianopsia, and aphasia, who also had a history of diabetes mellitus and chronic heart disease. **A** CT scan within 24 h after a major stroke, showing normal findings; **B** CT scan 10 days after the stroke, showing a large hypodensity area in the territory of the left anterior, middle, and posterior cerebral arteries; **C** ^{15}O studies within 24 h after the stroke, showing a marked reduction in rCBF associated with a marked increase of rOER, a reduction in rCBV, and a marked reduction in $rCMRO_2$ in the left posterior parietal region (*arrow*); **D** ^{18}F-FDG study within 44 h after the stroke, showing marked reduction of rCMRGl in the left parietal region (*arrow*); **E** ^{11}C-DMO study within 44 h after the stroke, showing a focal acidosis in the left parietal region (*arrow*)

87% reduction ($P < 0.005$). Values of rCpH in the infarcted area were lower than those in the contralateral homologous areas, suggesting an accumulation of acid metabolites. In the hyperglycemic cases (cases 5 and 7), rCpH values were more markedly reduced in the infarcted areas, indicating severe accumulation of acid metabolites. These findings are illustrated in Fig. 2A–E.

The clinical and laboratory investigations in this group indicated that cases 7 and 8 had a history of atrial fibrillation, and case 5 had a long history of chronic heart disease. Therefore, patients in this group could be at greater risk of embolic episodes than the patients in group I.

Discussion

Eleven patients underwent PET studies of six variables (rCBF, $rCMRO_2$, rOEF, rCBV, rCMRGl, and rCpH) within 48 h of the clinical onset of stroke. CT scans were also performed within 48 h, and again 1 week after clinical onset for anatomical confirmation of infarct localization.

In this study, we used functional values of the nonaffected contralateral homologous region as the reference to which we compared the focal functional changes in the affected hemisphere. This method of internal control is not ideal, since it does not take into account the phenomenon of diaschisis [3, 21, 27, 30], but it does decrease the influence of interpatient variability. We found that the area of focal hyperemia had significantly different metabolic patterns, as suggested by Lassen and his associates, who hypothesized that in "luxury perfusion" the accumulation of acid metabolites resulted in vasodilation and a varying degree of vasomotor paresis, depending on the degree of pH decrease [29, 42–44].

The metabolic characteristics of focal hyperemia (group I) and focal ischemia (group II) are shown in Fig. 3. The focal hyperemia group had a pattern of relative increase in rCBF associated with marked reduction (below 65 μmol/100 g/min) in $rCMRO_2$, indicating an irreversible

Fig. 3 Comparison of six metabolic parameters (rCBF, $rCMRO_2$, rOEF, rCBV, rCMRGl, and rCpH) between groups I and II, showing the differences between the metabolic parameters. *D*, diseased site; *C*, contralateral homologous site

process of cerebral damage, marked reduction in rOEF (a further indication of the cerebral infarcting process), and slightly higher rCpH values in the focal hyperemic areas as compared to the contralateral homologous regions. Our findings of metabolic changes in the focal hyperemic area, indicative of irreversible cerebral tissue damage, were further confirmed by a series of CT scans showing a zone of focal hypodensity. No evidence of focal acidosis was noted in any of the cases in group I.

In contrast, focal acidosis was present in all cases in group II, and was especially pronounced in patients with hyperglycemia (cases 5 and 7). Syrota et al. [47] reported that between 10 and 34 days after onset the infarcted areas showed alkalosis in all of the nine patients they studied. They suggested that the presence of a large extracellular fluid space or an increase in intracellular pH, or both, may be responsible for increased pH in the infarcted cerebral tissue. However, no local alkalosis was observed in the infarcted areas in group II. The focal ischemia associated with cerebral infarction in group II was associated with more acidosis than was present in the contralateral cerebral hemisphere. This could be due to inadequate clearance of acid metabolites, resulting from poor perfusion in the infarct area.

We observed the focal hyperemic phenomenon in the infarcted areas in 4 (36%) of 11 patients, an incidence similar to that in studies reported by Skyhoj Olsen et al. [42–44], using the ^{133}Xe intracarotid injection method. The focal hyperemia in the cerebral infarcting areas observed in our study showed a significant reduction in both rOEF and rCMRO$_2$, without any evidence of acidosis. These results suggested the presence of hyperoxic or arterialized "red cerebral veins," a phenomenon first described by Penfield [32, 33], who noted arterialized blood appearing locally in the pial vein after a focal cortical seizure or in veins draining cystic brain scars. The metabolic characteristics of focal hyperemia as studied here indicate the presence of metabolic shunts caused by a significant reduction in metabolizing tissue as a consequence of cerebral tissue damage. A general tendency, but at nonsignificant levels, to increased rCBV in the cerebral infarcts suggests the presence of mild vasodilation. This may not be entirely due to acidosis, since our rCpH studied over the infarcts show slightly more alkalosis than acidosis. Several vasoactive substances may be produced as a consequence of neuronal damage related to events disturbing intracellular ionic homeostasis, such as lipolysis and proteolysis [35, 36, 40, 41]. These products may be responsible for local alteration in vasomotor tone in the infarcting lesions.

For rCMRGl measurements with ^{18}F-FDG we had used the normal lumped and rate constant values reported by Huang et al. [25]. The rCMRGl values obtained in this study may therefore be considerably underestimated in the cerebral infarcting areas [16, 20]. However, we have observed a significant difference in the patterns of rCMRGl changes between group I, (where there were no significant changes in rCMRGl values as compared to the contralateral cerebral hemisphere), and group II, (where there was a significant reduction in rCMRGl values as compared to the contralateral cerebral hemisphere). Why glucose utilization was preserved in the area of focal hyperemia following cerebral tissue damage is difficult to explain. It may be related to abnormal metabolic processes, such as the removal of waste products and proliferation of glial cells [8, 23, 24] and/or infiltration by macrophagic cells [37]. However, how glucose utilization was preserved during the infarcting process despite the attendant local hyperemia remains to be determined.

In conclusion, two distinct metabolic patterns can be identified in the acute stages of stroke:

1. Focal hyperemia associated with slightly higher rCpH than in the contralateral homologous cerebral tissue (group I)
2. Focal ischemia associated with reduction in rCpH (group II)

Therefore, there was a correlation between rCBF changes and rCpH changes in the acute stages of stroke. However, no significant correlation was noted between rCpH changes and rOEF. Syrota et al. [47] did observe a correlation between rCpH and rOEF in chronic stroke. In the acute stages of stroke it appears that rOEF changes are time-related events, rather than true metabolic characteristics.

Acknowledgments. This work was supported by grant SP-5 from the Medical Research Council of Canada. We thank the staff of the PET unit for technical assistance, and Dr. V. Lees and Miss C. Elliot for their help in the preparation of this manuscript.

References

1. Ackerman RH, Correia JA, Alpert NM, Baron JC, Gouliamos A, Brownell GL, Taveras JM (1981) Positron imaging in ischemic stroke disease using compounds labelled with oxygen-15: initial results of clinicophysiologic correlations. Arch Neurol 38:537–543
2. Baron JC, Comar D, Bousser MG, Plummer D, Loch C, Kellershohn C, Castaigne P (1979) Patterns of CBF and oxygen extraction (EO_2) in hemispheric infarcts: A tomographic study with the ^{15}O continuous inhalation technique. Acta Neurol Scand 60 [Suppl 72]:324–325
3. Baron JC, Bousser MG, Comar D, Soussaline F, Castaigne P (1981) Non-invasive tomographic study of cerebral blood flow and oxygen metabolism in vivo. Eur Neurol 20:273–284
4. Bida GT, Satyamurthy N, Barrio JR (1984) The synthesis of 2[F-18]fluoro-2-deoxy-D-glucose using glycals: a re-examination. J Nucl Med 25:1327–1334
5. Brooks R (1982) Alternative formula for glucose utilization using labelled deoxyglucose. J Nucl Med 23:538–539
6. Cook BE, Evans AC (1983) A phantom to assess quantitative recovery in positron tomography. J Comput Assist Tomogr 7:876–880
7. Cook BE, Evans AC, Fanthome EC et al. (1984) Performance figures and images from the Therascan 3128 positron emission tomography. IEEE Trans Nucl Sci NS-31: 640–644
8. Cummins CJ, Lust WD, Passonneau JV (1983) Regulation of glycogen metabolism in primary and transformed astrocytes in vitro. J Neurochem 40:128–136
9. Diksic M (1984) A new, simple, high-yield synthesis of "no-carrier-added" ^{11}C-labelled DMO. Int J Appl Radiat Isot 35:1035–1037
10. Diksic M, Jolly D (1983) New high-yield synthesis of ^{18}F-labelled 2-deoxy-2-fluoro-D-glucose. Int J Appl Radiat Isot 34:894–896
11. Diksic M, Jolly D, Farrokhzad S (1982) An on-line synthesis of "no-carrier-added" ^{11}C-phosgene. Int J Nucl Med Biol 9:238–284
12. Feindel W, Perot P (1965) Red cerebral veins. J Neurosurg 22:315–325
13. Feindel W, Yamamoto YL (1967) Luxury — perfusion syndrome. Lancet I:48–49
14. Frackowiak RSJ, Lenzi GL, Jones T, Heather JD (1980) Quantitative measurement of regional cerebral blood flow and oxygen metabolism in man using ^{15}O and positron emission tomography: theory, procedure, and normal values. J Comput Assist Tomogr 4:727–736
15. Ginos JZ, Tilbury RS, Haber MT, Rottenberg DA (1982) Synthesis of 2-^{11}C-5,5-dimethyl-2,4-oxazolidinedione for studies with positron tomography. J Nucl Med 23:255–258
16. Greenberg JH, Reivich M (1983) Autoradiographic determination of local cerebral glucose metabolism: physiological and pathological studies. In: Szabo AJ (ed) Advances in metabolic disorders, vol 10. Academic, New York, pp 67–133
17. Hakim AM, Pokrupa R, Kitamura S, Evans A, Diksic M, Yamamoto YL, Feindel W (1984a) PET studies of circulation, energy metabolism and acid-base status after acute cerebral infarction in man. Nineteenth Canadian Congress of Neurological Sciences, Edmonton, Alberta
18. Hakim AM, Pokrupa R, Kitamura S, Yamamoto YL, Meyer E, Diksic M, Thompson CJ, Evans A, Feindel W (1984b) Acute cerebral infarction in humans: studies of cerebral circulation, metabolism and acid-base status by PET. 109th Annual Meeting of the American Neurological Association, Baltimore, Maryland, October 7–10

19. Hakim AM, Pokrupa RP, Diksic M, Evans AC, Thompson CJ, Meyer E, Yamamoto YL, Feindel W (1985) Acute cerebral infarction in man: studies of cerebral perfusion, metabolism and acid-base status by positron emission tomography. (In press)
20. Hawkins RA, Phelps ME, Huang SC, Kuhl DE (1981) Effect of ischemia on quantification of local cerebral glucose metabolic rate in man. J Cereb Blood Flow Metab 1:37–51
21. Heiss WD, Pawlik G, Wagner R, Ilsen HW, Herholz K, Wienhard K (1983) Functional hypometabolism in non-infarcted brain regions in ischemic stroke. J Cereb Blood Flow Metab 3 [Suppl 1]:582–583
22. Heiss WD, Wienhard K, Pawlik G, Wagner R, Ilsen HW, Herholz K (1985) Hypometabolism in stroke: cerebral metabolic rate for glucose in infarcted and remote tissue obtained by dynamic determination of individual kinetic constants. In: Greitz T, Ingvar DH, Widén L(eds) The metabolism of the human brain studied with positron emission tomography, Raven, New York, pp 399–410
23. Hertz L (1979) Functional interactions between neurons and astrocytes: turnover and metabolism of putative amino acid transmitters. Prog Neurobiol 13:277–323
24. Hertz L (1981) Features of astrocytic function apparently involved in the response of central nervous tissue to ischemic hypoxia. J Cereb Blood Flow Metab 1:143–153
25. Huang SC, Phelps M, Hoffman E, Sideris K, Selin CJ, Kuhl DE (1980) Non-invasive determination of local cerebral metabolic rate of glucose in man. Am J Physiol 238: 69–82
26. Kearfott KJ, Junck L, Rottenberg DA (1983) ^{11}C-dimethyloxazolidinedione (DMO): Biodistribution, estimates of radiation absorbed dose and potential for positron emission tomography (PET) measurement of regional brain tissue pH. J Nucl Med 24:805–811
27. Kuhl DE, Phelps ME, Kornwell RP, Metter EJ, Seline C, Winter J (1980) Effects of stroke on local cerebral metabolism and perfusion: mapping by emission computed tomography of ^{18}FDG and $^{13}NH_3$. Ann Neurol 8:47–60
28. Lammertsma AA, Jones T, Frackowiak RSJ, Lenzi GL (1981) A theoretical study of the steady-state model for measuring regional cerebral blood flow and oxygen utilization using oxygen-15. J Comput Assist Tomogr 5:544–550
29. Lassen NA (1966) The luxury-perfusion syndrome and its possible relation to acute metabolic acidosis localized within the brain. Lancet II:1113–1115
30. Lenzi GL, Frackowiak RSJ, Jones T et al. (1981) $CMRO_2$ and CBF by oxygen-15 inhalation technique. Results in normal volunteers and cerebrovascular patients. Eur Neurol 20:285–290
31. Lenzi GL, Frackowiak RSJ, Jones T (1982) Cerebral oxygen metabolism and blood flow in human cerebral ischemic infarction. J Cereb Blood Flow Metab 2:321–335
32. Penfield W (1933) The evidence for a cerebral vascular mechanism in epilepsy. Ann Intern Med 7:303–310
33. Penfield W (1938) The circulation of the epileptic brain. Res Publ Assoc Res Nerv Ment Disc 18:605–637
34. Phelps ME, Huang SC, Hoffman E, Kuhl DE (1979) Tomographic measurement of local cerebral glucose metabolic rate in humans with ^{18}F-fluoro-2-deoxy-D-glucose: validation of method. Ann Neurol 6:371–388
35. Raichle ME (1983) The pathophysiology of brain ischemia. Ann Neurol 13:2–10
36. Rehncrona S, Westerberg E, Akesson B, Siesjö BK (1982) Brain cortical fatty acids and phospholipids during and following complete and severe incomplete ischemia. J Neurochem 38:84–93

37. Rhodes CG, Wise RJS, Hatazawa J, Frackowiak RSJ, Palmer AJ, Jones T (1982) Mismatching between cerebral oxygen and glucose metabolism in patients with cerebral glioma and stroke. In: Raynaud (ed) Nuclear medicine and biology, Proceedings of Third World Congress, Paris, vol 2. Pergamon, Paris, pp 2200–2203
38. Robinson GD Jr (1984) private communication
39. Rottenberg DA, Ginos JZ, Kearfott KJ, Junck L et al. (1985) In vivo measurement of brain tumor pH using ^{11}C-DMO and positron emission tomography. Ann Neurol 17:70–79
40. Samuelsson B, Borgeat P, Hammarstrom S et al. (1980) Leukotrienes: a new group of biologically active compounds. Adv Prostoglandin Thromboxane Leukotriene Res 6:1–18
41. Siesjö BK (1984) Cerebral circulation and metabolism. J Neurosurg 60:883–908
42. Skyhoj Olsen T, Larsen B, Skriver EB, Herning M, Enevoldsen E, Lassen NA (1981) Focal cerebral hyperemia in acute stroke; incidence, pathophysiology and clinical significance. Stroke 12:598–606
43. Skyhoj Olsen T, Larsen B, Herning M, Skriver EB, Lassen NA (1983) Blood flow and vascular reactivity in collaterally perfused brain tissue. Stroke 14:332–341
44. Skyhoj Olsen T, Lassen NA (1984) A dynamic concept of middle cerebral artery occlusion and cerebral infarction in the acute state based on interpreting severe hyperemia as a sign of embolic migration. Stroke 15:458–468
45. Sokoloff L, Reivich M, Kennedy C et al. (1977) The ^{14}C-deoxyglucose method for the measurement of local cerebral glucose utilization: theory, procedure, and normal values in the conscious and anesthetized albino rat. J Neurochem 28:897–916
46. Strother SC, Evans AC, Thompson CJ (1984) Testing quantitation in PET. J Nucl Med 25:107
47. Syrota A, Castaing M, Rougemont D, Berridge M, Maziere B, Baron JC, Bousser MG, Rocidalo JJ (1983) Tissue acid base balance and oxygen metabolism in human cerebral infarction studied with positron emission tomography. Ann Neurol 14:419–428
48. Wise RJS, Bernardi S, Frackowiak RSJ, Legg NJ, Jones T (1983a) Serial observations on the pathophysiology of acute stroke. Brain 106:197–222
49. Wise RJS, Rhodes CG, Gibbs JM, Hatazawa J, Palmer T, Frackowiak RSJ, Jones T (1983b) Disturbance of oxidative metabolism of glucose in recent human cerebral infarcts. Ann Neurol 14:627–637
50. Yamaguchi T, Waltz AG, Okazaki H (1971) Hyperemia and ischemia in experimental cerebral infarction: correlation of histopathology and regional blood flow. Neurology (Minneap) 21:565–578
51. Yamamoto YL, Phillips KM, Hodge CP, Feindel W (1971) Microregional blood flow changes in experimental cerebral ischemia: effects of arterial carbon dioxide studies by fluorescein angiography and xenon-133 clearance. J Neurosurg 35:155–166

Measurement of Cerebral Protein Synthesis and Intracellular pH in Man by Positron Emission Tomography

A. SYROTA

Positron emission tomography (PET) has provided a method for the noninvasive study of brain function. Measurement of regional cerebral blood flow (rCBF), and of oxygen and glucose metabolism are now widely used in clinical conditions. More recently, methods for measuring protein synthesis and intracellular pH have also been reported.

Investigation of Brain Intracellular pH with PET in Stroke Patients

A large number of hemodynamic and biochemical alterations of the brain during ischemia have been produced in animals using various experimental stroke models [70]. However, these abnormalities have usually been identified during or immediately after the interruption of CBF in experiments that measured only a limited number of physiological variables. The coupling between blood flow, oxygen metabolism, glucose metabolism, regulation of cell volume, and pH is altered in ischemic stroke models [31, 57]. Thus, it seemed important to study the nature of these alterations in patients with recent cerebral infarction.

Such an investigation, however, requires a noninvasive methodology. So far, PET is the only one which offers the possibility of studying rCBF and regional oxygen and glucose utilization noninvasively in human cerebral ischemic infarction [6, 72]. Because changes in intracellular pH are an important consequence of alterations in cell metabolism, the application of PET in the measurement of brain pH has recently been proposed using either DMO (5,5-dimethyl-2,4-oxazolidinedione) [54, 65] or carbon dioxide [13, 18, 49]. These weak acids were labelled with the short half-lived positron emitter ^{11}C, and PET scans were performed at fixed axial tomographic level.

Measurement of Regional Cerebral Intracellular pH by PET

Two methods have been developed for measuring intracellular pH (pH_i) by PET. Both techniques are based on the use of a weak acid labelled with ^{11}C. After injection, the weak acid AH dissociates to give:

$$AH \leftrightharpoons A^- + H^+$$

Usually, the cell membranes are freely permeable to the unionized form AH and impermeable to the ionized form A^-, so that the ratio of ionized to unionized AH in any compartment is related to the pH prevailing in this compartment, according to the Henderson-Hasselbalch equation. Measurement of regional cerebral pH_i by PET requires in addition the measurement of total brain water volume and of extracellular water volume, since the weak acid diffuses into several compartments: blood, extracellular water, and intracellular water.

The weak acid must not be toxic and must not be metabolized. It must not bind to necrotic tissue and it must equilibrate in a time suitable for PET study. One must at least know how to label the molecule with a positron emitter. Two molecules have been used in man, namely $^{11}CO_2$ and ^{11}C-DMO.

A few years ago, Raichle [49] suggested that $^{11}CO_2$ could be used to study brain pH. Intravenous injections of $^{11}CO_2$ were given to monkeys, and measurements were made in blood and brain. However, Lockwood and Finn [35] found that, following a single-breath inhalation of $^{11}CO_2$ in rats, nearly 20% of the label was metabolized after 15 min. Buxton et al. [18] developed a kinetic model of the time-dependent tissue concentration of ^{11}C following administration of $^{11}CO_2$. This model was tested by Brooks et al. [13] in a few patients.

The DMO method for measuring intracellular pH has been amply discussed [52, 53, 69]. Briefly, the interest of the method for human studies relied on the fact that DMO is a nontoxic molecule, is not metabolized, equilibrates in a time suitable for PET study, and can be labelled with ^{11}C, a 20-min half-life positron emitter [12]. Between 20 and 50 min after injection, a DMO equilibrium distribution was achieved in total tissue water, constituting a closed system [65].

The brain pH could be calculated in each pixel of the PET scan, because the fractional volume of extracellular water was measured by means of an anion, ^{76}Br, which is also a positron emitter [65a].

Changes in Extracellular Water Volume and pH_i in Recent Cerebral Infarction

The size of the brain extracellular water volume has been determined in various animal models of cerebral ischemia [27, 43]. Regional measurement of the fraction of extracellular water volume (FECWV) is made possible in man by PET because the radioactive concentration of a tracer can be accurately measured in a defined volume selected in a tomographic slice. Many substances have been assessed for the determination of the extracellular fluid volume in animals. They can be classified in two groups: those having a volume of distribution similar to that of inulin, which include mannitol, raffinose, thiosulfate, and sulfate, and those with a volume of distribution similar to chloride, which include sodium, thiocyanate, and bromide. A recent comparative analysis showed that the steady-state spaces of ^{3}H-raffinose, ^{3}H-inulin, and ^{14}C-inulin were not significantly different; however, the ^{3}H-mannitol, ^{22}Na, and ^{36}Cl spaces were greater than that of ^{3}H-inulin [59]. The exact magnitude of extracellular volume (ECV) in brain is not known. However, the distribution volume of inulin may underestimate ECV, since this marker is not distributed over the entire connective tissue water, whereas the distribution space of chloride or bromide tends to overrate ECV, because these ions migrate into cells and are lost via the gastrointestinal tract [62].

In a recent study, Syrota et al. [65 a] found that the FECWV was 12% of total brain water volume in the normal hemisphere of the patients. The dispersion of the values around this average within and between patients was very low. The FECWV value measured in each region of interest (ROI) was a weighted value representing extracellular space of both gray and white matter, since an ROI had an area of 2.9 cm^2 and a thickness of 1.9 cm; the ratio of gray to white matter in each ROI was unknown, but the regions were mainly selected over the cortex, so they could represent a higher percentage of gray than of white matter. The values obtained in animals differed according to method and species, but were not very different from those obtained by PET. Pelligrino et al. [46] measured the sulfate space in the dog brain and reported values of 22% of tissue weight in cortical gray matter and 5% in periventricular white matter. Arieff et al. [2] found a mean sulfate space of 22%; Kibler et al. [28] found a value of only 3% with the same marker, but a much higher one, 39%, for the mean chloride space. Other authors recorded the cortical impedance to measure the volume fraction of normal cortical extracellular space, and reported values ranging from 15% to 24% [21, 24, 39].

Brain pH_i has often been measured in animals with ^{14}C-DMO as a marker; Kibler [28] reported a value of 6.69 in the dog when chloride space was measured, and 7.09 using sulfate space. Arieff [2] found values of 7.05 for cerebral cortex and 6.97 for white matter, and Pelligrino [46] 7.07 for cortical gray matter and 7.12 for periventricular white matter, also in the dog brain. Using the same DMO method, Roos recorded a value of 7.13 for cat brain pH_i (1965). Using ^{31}P NMR spectroscopy, pH_i was found to be 7.20 in the gerbil brain [66] and 7.27 in the rat brain [42]. Anderson [1] reported a value of 7.0 in the squirrel monkey with a pH-sensitive fluorescent indicator.

Human brain pH values have very recently been published. Rottenberg et al. [54] used the ^{11}C-DMO method, but did not measure the size of the extracellular space, and their values represent a weighted average of intra- and extracellular pH. Syrota et al. [65a] calculated a true intracellular pH in ten patients, and found a value of 6.86 ± 0.11. This mean pH_i value represented a weighted average of the pH_i values of all cells within a resolution element of the positron camera: neuronal, glial, and possibly inflammatory cells. Brooks et al. [13] used the $^{11}CO_2$ continuous inhalation technique to measure regional cerebral pH. The FECWV was not measured, but using an FECWV value of 20% and an extracellular pH of 7.40 Brooks et al. [13] calculated an intracerebral pH around 6.77. Using ^{31}P-NMR [19], Hope et al., [23] reported a value of 7.14 in infants.

In recent cerebral infarction (patients seen between 10 and 19 days after onset of symptoms), Syrota et al. [65a] found that the extracellular water volume was increased in the infarcted area: in two patients FECWV represented 50% of total brain water, instead of 12% in the normal tissue. Use of the bromide space as a measure of the extracellular fluid volume in the infarcted area could lead to an overestimate, because ^{76}Br could enter necrotic cells. Bromine is excluded from the intracellular fluid by the transmembrane potential; in the absence of an energy source, this potential will decrease and the intracellular concentration of chloride and bromide will increase. The extracellular water volume measured in the infarct would thus represent the difference between the total water volume and intracellular water volume of living cells.

The main result concerning the changes in pH_i in recent cerebral infarction was the finding of an intracellular alkalosis which was demonstrated in the infarcted tissue of seven out of ten patients [65 a]. In all the patients cerebral blood flow (CBF), oxygen extraction fraction (OEF), and oxygen consumption ($CMRO_2$) were also measured using the $^{15}O_2/C^{15}O_2$ continuous inhalation technique [26, 34].

The change in pH_i was neither correlated to the change in FECWV, nor to those in CBF or $CMRO_2$. An increase of 0.20 pH u was observed in two patients, one of whom had a 40% increase in CBF, the other a 30% decrease. However, a significant correlation was observed between changes in pH_i and in OEF; the higher the pH_i, the lower the OEF in the infarcted area. This means that there is a relationship between local luxury perfusion and intracellular pH. In infarcted and ischemic areas the aerobic metabolism is reduced, as seen from the decreased oxygen consumption in all patients. Therefore, CO_2 and other acid metabolites are produced in smaller amounts in the ischemic tissue and are removed by local perfusion if they are in excess of metabolic needs [65]. As a result, the H^+ activity will be lower than normal, leading to local alkalosis. This does not preclude the existence of tissue acidosis within the first few hours after onset of symptoms, since our patients were seen between 10 and 19 days after onset.

Mabe et al. [37] demonstrated an intracellular alkalosis, following recirculation to areas of brain which had been ischemic for 15 min in rats. The alkalosis was still present 90 min after the start of recirculation. Kogure et al. [31] also found that in the rat brain tissue became alkaline 30 min after embolization. Quantitative autoradiography with ^{14}C-DMO was recently used to study the local changes in brain pH after ischemia in the rat [30]. The authors did not measure the extracellular water space and assumed that it occupied 15% of gray matter; they reported a reduced ^{14}C-DMO concentration after middle cerebral artery (MCA) occlusion in the areas coinciding with the most severe reduction in CBF. However, no measurement was made more than 4 h after MCA occlusion.

The presence of local alkalosis in recent cerebral infarction may have important consequences for brain glucose metabolism. Shifts in tissue pH affect the glycolysis rate by changing the affinity of phosphofructokinase (EC 2.7.1.11) for fructose-6-phosphate [36]; glycolysis is promoted by increased activity of phosphofructokinase at an alkaline pH [67, 68]. This hypothesis was confirmed in two recent publications. Wise et al. [72] studied eight patients with recent cerebral hemispheric infarction with PET and the $^{15}O_2$ inhalation and ^{18}F-deoxyglucose (FDG) techniques. They found that both regional oxygen consumption ($rCMRO_2$) and glucose metabolism (rCMRGl) were significantly depressed within the infarct. The OEF was also low, and the metabolic ratio (the ratio of $rCMRO_2$ to rCMRGl) was reduced to 0.25 ml O_2/mg glucose instead of 0.68–0.75 in the normal brain; however, the glucose extraction fraction was increased. Baron et al. [7] also found an uncoupling between $CMRO_2$ and CMRGl in all seven stroke patients studied. A depressed metabolic rate was found in the necrotic core of two recent stroke cases, together with a reduction in OEF.

Thus, these studies indicate that in the infarcted tissue of patients with recent infarct, oxygen consumption is not limited by oxygen supply, since OEF is generally low and glucose metabolism is much less affected than oxygen metabolism, indicating nonoxidative utilization of glucose. Furthermore, as reported in our study, intracellular pH is shifted to alkaline pH. Wise et al. [72] have suggested that an aerobic production of lactate (aerobic glycolysis) could be attributable to the presence of numerous active phagocytic cells, which are known to infiltrate the ischemic tissue within 24 h of the onset of ischemia [73].

In neutrophils, glycolysis is the predominant mechanism for energy production [63]. Beck [9] and Sbarra et al. [56] demonstrated that leukocytes obtained from human blood show a very high aerobic glycolysis, and that about 20 times more glucose carbon is metabolized to lactate than is converted to CO_2. Glycolysis under both aerobic and anaerobic conditions is essential for phagocytosis; the burst of oxygen consumption observed during phagocytosis is related to the synthesis of hydrogen peroxide catalyzed by the enzyme superoxide dismutase, and not to the oxidation of glucose [3]. Thus, one might speculate that the luxury perfusion resulting from revascularization in postischemic brain could explain the alkaline shift in pH_i, which in turn activates the glycolysis necessary for the phagocytosis of necrotic cells by inflammatory cells and favors the dismutation reaction with production of superoxide anion [29].

Measurement of Cerebral Protein Synthesis by PET

Choise of Biochemical Probe

Optimal Selection of a Tracer is Based on Three Points:

Biochemical parameters. The tracer must have a rapid incorporation in proteins. It must be absorbed noticeably by the organ under study. Its intracellular neosynthesis and recycling should be zero. A choice must therefore be made from the eight amino acids essential for the human species. Nonessential amino acids are strongly synthesized by cells and also, with regard to cellular energy saving, actively reused in situ after protein catabolism. Essential amino acids not synthesized by cells are taken from the intercellular fluid and charged on tRNA within a few seconds [50]. Their intracellular pool is very small and is always located on the degradation side of protein metabolism, without recycling toward protein synthesis in the cytoplasm [71]. Incorporation of such a tracer in proteins is also very rapid, linear as early as the first few seconds, and dependent on the extracellular presence of the tracer [71].

Brain amino acid uptake is particularly well documented in the literature, and the brain uptake index (B. U. I.) method [44, 45] has been used to study the amino acid transport systems. B. U. I. values vary from 55% for phenylalanine in perfusion media without proteins to less than 12% for threonine. These values are reduced to 22% and 4%, respectively, when the amino acids are injected into plasma. These values are 3–4 times higher than those of nonessential amino acids and, therefore, interesting for in vivo studies. Methionine has a B.U.I. of 36% in synthetic saline serum, but of only 13%–18% when it competes with normal levels of other large neutral amino acids crossing the blood–brain barrier (BBB) by the same L-system.

Arteriovenous differences must also be taken into account: a negative value means a *de novo* synthesis of the amino acid, which is then preferentially used by cells. However, the amino acid coming from the extracellular space is the first and often the only one to be marked after an intravenous injection of the radioactive amino acid in blood. If a negative arteriovenous difference is a good criterion for excluding an amino acid, a positive or zero arteriovenous difference will not exclude in situ recycling or de novo synthesis. According to this cirterion, the best candidates for the determination of protein synthesis are L-methionine, L-leucine, L-valine, and L-phenylalanine, which are all essential amino acids.

Another essential criterion is an as much as possible unbranched metabolism. The relative values of fluxes between protein synthesis and catabolic needs of the tissue can vary widely in pathologic conditions. It could be dangerous to assume them to be constant and to include them in the hypotheses of a physiological quantitative model for protein synthesis. This would severely limit the clinical applications of the method.

Furthermore, the complexity of the mathematical model needed to fit raw data increases with the number of pathways describing the metabolism of the tracer, and rapidly exceeds the possibilities of compartmental analysis or yields confidence limits which are too large and nonsignificant. However, if protein synthesis is the main fate of the tracer, a wrong estimation of the percentage of a secondary metabolism of the active part of the molecule will have minimal effect.

Plasmatic stability of the tracer is the last important point to check. A possible transformation of the tracer by peripheral organs (muscles or liver during interprandial conditions) may occur.

For the above reasons, L-methionine is a good candidate for the study of protein synthesis in brain tissue, because it has a high B. U. I., one of the most marked arteriovenous differences [5]. Furthermore, L-methionine has a small metabolism toward the transmethylation pathway [32, 61, 64].

Radiochemical Parameters. In spite of the fact that the metabolism of amino acids is rather "slow" compared to oxidative metabolism, only short half-lived positron-emitting isotopes can be used for the in vivo tracing of protein synthesis by PET, namely: ^{13}N, ^{11}C, and ^{18}F.

The Sokoloff autoradiographic method [60] permits the collection of data only after the disappearance of the exchangeable labelled amino acid in plasma and brain, and this condition does not occur less than 70 min after i. v. injection. This is a factor which limits the use of amino acids labelled with ^{11}C.

Many enzymatic methods for labelling amino acids by labelling the amine group with ^{13}N have been published, but no model of the incorporation in protein has yet been constructed, and the half-life of ^{13}N is too short. Fluorination by ^{18}F of the aromatic nucleus of phenylalanine or dopa could be interesting because of the longer half-life of ^{18}F, but the kinetic parameters characterizing the transport of the fluoro-amino acid are different from those for the transport of native amino acid. Labelling with ^{11}C on the carboxylic group is widely used. Racemization of the product introduces a difficult step for the separation of L and D forms, the former being the only metabolized one. This step is generally performed by chromatography. A rapid method for isolation of the pure L-amino acid isomer from the enantiomeric adsorption of D-amino acid oxidase and catalase has been published [8].

Methyl groups, such as in L-methionine, are easily labelled with ^{11}C: $^{11}CO_2$, produced by the cyclotron, is transformed into $^{11}CH_3$, which reacts on L-homocysteine to give ^{11}C-L-methionine [10, 20]. After a 10-min irradiation and 20–30-min chemical workup, including chromatographic purification, it is possible to routinely obtain more than 200 mCi of ^{11}C-L-methionine ready for injection, with a purity higher than 99% and a specific activity higher than 600 Ci/mmol [20].

Physical Constraints. The main constraint is the short half-life of positron emitters. This parameter will govern the optimum time of measurement, taking into account the rate of incorporation of the amino acid into proteins, the amount of tracer injected, and dosimetry. Study of the sensitivity of the kinetic model described below shows that optimal sampling time is around 40–50 min after injection of ^{11}C-L-methionine; at that time brain activity becomes constant.

Experimental Protocol and Modelling

^{11}C-L-methionine Model. Based on preliminary work using autoradiography, this model was validated for the quantification of protein synthesis in rats and baboons, and finally adapted for PET measurement in humans [14]. Biochemical basis, tracer localization in neurons, etc., are particularly well described in the literature, since ^{3}H and ^{35}S-methionine have been extensively used to follow in vivo protein synthesis [22, 32, 40].

A three-compartment model is theoretically fully defined if one can measure the input function (plasma activity of exchangeable tracer) and the change in brain radioactivity with time. The model provides enough information to correctly and independently identify three (but only three) transport rate constants. The protocol included two steps. A study of regional cerebral blood volume (rCBV) was first performed using ^{68}Ga; ^{11}C-L-methionine was then injected, and PET scans were performed at the same brain level. Subtraction of blood activity from ^{11}C-methionine scans was performed pixel by pixel; from the free exchangeable methionine activity curve, each elementary brain activity curve was fitted to the model. Values of regional kinetic parameters were then calculated and used to produce, pixel by pixel, a tomographic mapping of local methionine input, the half-life of free methionine, the partition coefficient between blood and intercellular compartment at the BBB level, and the local protein synthesis rate. Direct validation of the model was obtained from baboons [15].

$^{11}C_1$-L-leucine Model. This model was proposed by Smith [58] after autoradiographic validation, and adapted by Phelps [47, 48] for human studies. It is necessary to consider the important catabolism of leucine and its precursor pool, which is distinct from that leading to protein synthesis. The key position of C_1-labelling in the carboxyl group allows the escape of ^{11}C from leucine shortly after it enters catabolism [4]. The diffusion of produced $^{11}CO_2$ from brain toward blood and lungs is very rapid. Experimental data can be fitted by means of a four-compartment model with two pools of free amino acid, only one being concerned with protein synthesis. The four transport rate constants cannot be calculated independently, and one hypothesis must relate two of the constants, or their ratio has to be fixed. Phelps et al. [48] published values of ^{11}C-leucine metabolism in baboons and man which were in agreement with those obtained by Bustany et al. [14, 15, 16] with ^{11}C-L-methionine.

A quantitative autoradiographic model using valine was described in rats [20a]. It implied the injection of considerable mass of labelled valine with a low specific activity in order to saturate the external precursor pool of protein synthesis. It was then possible to consider plasma-exchangeable valine as being the direct precursor pool of protein synthesis. This method has not been validated for PET studies so far.

Protein Synthesis in Clinical Conditions

Protein Synthesis in Alzheimer-type Dementia. Kinetic studies of protein synthesis were performed in a group of senilely demented Alzheimer-type patients (SDAT) [17]. All the CT scans were normal, compared to a cross-matched age reference group. A decrease in protein synthesis without BBB alteration was found in all subjects, and was related to the severity of the disease as reflected by Rey's memory test. At early stages in the disease (often only transient memory impairment), a 20% decrease in protein synthesis was found in frontal lobes. In more severely demented patients, a 75% decrease in protein synthesis was observed without any apparent modification in brain tissue density as shown on CT scans. Alteration in protein synthesis was seen mainly in frontal and temporal regions. Of newly synthetized proteins in brain 75% are later found in membranal and synaptic proteins of dendrites. This fact correlates with the biopsy findings in SDAT patients: a significant decrease in RNA concentration or tubule production for axonal transport, a reduction in acetylcholine transferase activity, etc. [38].

Impairment in protein synthesis before any significant psychometric or clinical abnormality emerges might suggest that PET study of local protein synthesis could be a sensitive method of evaluating neuronal metabolic alteration at early stages of Alzheimer's disease.

Brain Tumors. In low-grade astrocytomas, an increase in local protein synthesis of up to 450% was observed with the ^{11}C-L-methionine method. BBB transport rate constants were also increased, with a marked decrease in the local half-life of free methionine, owing to the high metabolic activity of the tumor. These studies confirmed those of Hubner et al. [25], using synthetic ^{11}C-amino acids, and the observations of Bergström et al. [11]. Very recently, Meyer et al. [41] reported that brain tumors accumulated both ^{11}C-L- and D-methionine, with the strongest uptake occurring in tumors with a high grade of malignancy.

References

1. Anderson RE, Sundt TM (1983) Brain pH in focal cerebral ischemia and the protective effect of barbiturate anesthesia. J Cereb Blood Flow Metab 3:493–497
2. Arieff AI, Kerian A, Massry SG, Delima J (1976) Intracellular pH of brain: alterations in acute respiratory acidosis and alkalosis. Am J Physiol 230:804–812
3. Badway JA, Karnovsky ML (1980) Active oxygen species and the functions of phagocytic leukocytes. Annu Rev Biochem 49:695–726
4. Banker G, Cotman CW (1971) Characteristics of different amino acids as protein precursors in mouse brain: advantages of certain carboxyl-labeled amino acids. Arch Biochem Biophys 142:565–573
5. Barak AJ, Beckenhauer HC, Tuma DJ (1982) Use of S-adenosylmethionine as an index of methionine recycling in rat liver slices. Anal Biochem 127:372–375
6. Baron JC, Lebrun-Grandié P, Collard P, et al. (1982) Noninvasive measurement of blood flow, oxygen consumption and glucose utilization in the same brain regions in man by positron emission tomography. J Nucl Med 23:391–399
7. Baron JC, Rougemont D, Soussaline F, et al. (1984) Local interrelationships of cerebral oxygen consumption and glucose utilization in normal subjects and in ischemic stroke patients: a positron tomographic study. J Cereb Blood Flow Metab 4:140–149
8. Barrio JR, Keen RE, Ropchan JR, MacDonald NS, Baumgartner FJ, Padgett HC, Phelps ME (1982) L-|1-^{11}C|Leucine: Routine synthesis by enzymatic resolution. J Nucl Med 24: 515–521
9. Beck WS (1958) The control of leukocyte glycolysis. J Biol Chem 232:251–270
10. Berger G, Mazière M, Knipper R, Prenant C, Comar D (1979) Automated synthesis of ^{11}C-labelled radiopharmaceuticals imipramine, chlorpromazine, nicotine and methionine. Int J Appl Radiat Isot 30:393–399
11. Bergström M, Collins VP, Ehrin E, et al. (1983) Discrepancies in brain tumor extent as shown by computed tomography and positron emission tomography using |^{68}Ga| EDTA, |^{11}C| glucose, and |^{11}C| methionine. J Comput Assist Tomogr 7:1062–1066
12. Berridge M, Comar D, Roeda D, Syrota A (1982) Synthesis and in vivo characteristics of |2-^{11}C| 5,5-dimethyloxazolidine 2,4-dione (DMO). Int J Appl Radiat Isot 33:647–651
13. Brooks DJ, Lammertsma AA, Beaney RP, et al. (1984) Measurement of regional cerebral pH in human subjects using continuous inhalation of $^{11}CO_2$ and positron emission tomography. J Cereb Blood Flow Metab 4:458–465
14. Bustany P, Sargent T, Saudubray JM, Henry JF, Comar D (1981) Regional human brain uptake and protein incorporation of ^{11}C-L-methionine studied in vivo with PET. J Cereb Blood Flow Metab 1:517–518
15. Bustany P, Henry JF, Sargent T, Zarifian E, Cabanis E, Collard P, Comar D (1983) Local brain protein metabolism in dementia and schizophrenia: in vivo studies with ^{11}C-L-methionine and positron emission tomography. In: Heiss WD, Phelps ME (eds) Positron emission tomography of the brain. Springer, Berlin Heidelberg New York, pp 208–211
16. Bustany P, Henry JF, Soussaline F, Comar D (1983) Brain protein synthesis in normal and demented patients – a study by positron emission tomography with ^{11}C-L-methionine. In: Magistretti P (ed) Functional radionuclide imaging of the brain. Raven, New York, pp 319–326
17. Bustany P, Henry JF, De Rotrou J, Signoret P, Cabanis E, Zarifian E, Ziegler M, Derlon JM, Crouzel C, Soussaline F, Comar D (1985) Correlations between clinical

state and PET measurement of local brain protein synthesis in Alzheimer's dementia, Parkinson's disease, schizophrenia and gliomas. In: Widén L (ed) The metabolism of the human brain studied with positron emission tomography. Raven, New York, pp 241–249
18. Buxton RB, Wechsler LR, Alpert NM, et al. (1984) Measurement of brain pH using $^{11}CO_2$ and positron emission tomography. J Cereb Blood Flow Metab 4:8–16
19. Cady EB, Costello AMDeL, Dawson MJ, et al. (1983) Noninvasive investigation of cerebral metabolism in newborn infants by phosphorus nuclear magnetic resonance spectroscopy. Lancet I:1059–1062
20. Comar D, Cartron JC, Mazière M, Marazano C (1976) Labelling and metabolism of methionine — methyl — ^{11}C. Eur J Nucl Med 1:11–14
20a. Dwyer BE, Donatoni P, Wasterlain CG (1982) A quantitative autoradiographic method for the measurement of local rates of brain protein synthesis. Neurochem Res 7:563–576
21. Fenstermacher JD, Li CL, Levin VA (1970) Extracellular space of the cerebral cortex of normothermic and hypothermic cats. Exp Neurol 27:101–114
22. Fischer J, Kolousek J, Lodin Z (1956) Incorporation of methionine (sulphur 35) into the central nervous system. Nature 178:1122–1123
23. Hope PL, Cady EB, Tofts PS, et al. (1984) Cerebral energy metabolism studied with phosphorus NMR spectroscopy in normal and birth-asphyxiated infants. Lancet II:366–369
24. Hossmann KA (1971) Cortical steady potential impedance and excitability changes during and after total ischemia of cat brain. Exp Neurol 32: 163–175
25. Hubner KF, Purvis JT, Mahaley SM, Robertson JT, Rogers S, Gibbs WD, King P, Partain CL (1982) Brain tumor imaging by positron emission computed tomography using ^{11}C-labeled amino acids. J Comput Assist Tomogr 6:544–550
26. Jones T, Chesler DA, Ter-Pogossian MM (1976) The continuous inhalation of oxygen-15 for assessing regional oxygen extraction in the brain of man. Br J Radiol 49:339–343
27. Katzman R, Clasen R, Klatzo I, et al. (1977) IV. Brain edema in stroke. Report of Joint Committee for Stroke Resources. 8:512–540
28. Kibler RF, O'Neill RP, Robin ED (1964) Intracellular acid-base relations of dog brain with reference to the brain extracellular volume. J Clin Invest 43:431–443
29. Klebanoff SJ (1980) Oxygen metabolism and the toxic properties of phagocytes. Ann Intern Med 93:480–489
30. Kobatake K, Sako K, Izawa M, et al. (1984) Autoradiographic determination of brain pH following middle cerebral artery occlusion in the rat. Stroke 15: 540–547
31. Kogure K, Busto R, Schwartzmann RJ, Scheinberg P (1980) The dissociation of cerebral blood flow, metabolism, and function in the early stages of developing cerebral infarction. Ann Neurol 8: 278–290
32. Kozik M, Ozarewska E (1974) Autoradiographic studies on methionine-^{3}H incorporation into rat brain. Exp Pathol 9:274–279
33. Langer BW (1964) Organ and intracellular location of the methionine methyl group synthesizing system of the rat. Proc Soc Exp Biol Med 115:1088–1090
34. Lebrun-Grandié P, Baron JC, Soussaline F, Loc'h C, Sastre J, Bousser MG (1983) The coupling between regional blood flow and oxygen utilization in the normal human brain: a study with positron tomography and oxygen 15. Arch Neurol 40:230–236
35. Lockwood AH, Finn RD (1982) ^{11}C-carbon dioxide fixation and equilibration in rat brain: effects on acid-base measurements. Neurology 32:451–454
36. Lowry OH, Passonneau JV, Hasselberger FY, Schulz DW (1964) Effects of ischemia on known substrates and cofactors of the glycolytic pathway in brain. J Biol Chem 239:18–30

37. Mabe H, Blomqvist P, Siesjö BK (1983) Intracellular pH in the brain following transient ischemia. J Cereb Blood Flow Metab 3:109–114
38. Mann DMA, Neary D, Yates PO, Lincoln J, Snowden JS, Stanworth P (1981) Alterations in protein synthetic capability of nerve cells in Alzheimer's disease. J Neurol Neurosurg Psychiatry 44:97–102
39. Matsuoka Y, Hossman KA (1982) Cortical impedance and extracellular volume changes following middle cerebral artery occlusion in cats. J Cereb Blood Flow Metab 2:466–474
40. Merei FT, Gallyas F (1964) Quantitative determination of the uptake of (^{35}S) methionine in different regions of the normal rat brain. J Neurochem 11:257–264
41. Meyer GJ, Schober O, Hundeshagen H (1985) Uptake of ^{11}C-L- and D-methionine in brain tumors. Eur J Nucl Med (In Press)
42. Naruse S, Takada S, Koizuka I, Watari H (1983) In vivo ^{31}P NMR studies on experimental cerebral infarction. Jpn J Physiol 33:19–28
43. O'Brien MD, Waltz AG, Jordan M (1974) Ischemic cerebral edema. Distribution of water in brains of cats after occlusion of the middle cerebral artery. Arch Neurol 30:456–460
44. Oldendorf WH (1971) Uptake of radiolabeled essential amino acids by brain following arterial injection. Proc Soc Exp Biol Med 13:385–386
45. Oldendorf WH (1981) Clearance of radiolabeled substances by brain after arterial injection using a diffusible internal standard. In: Marks N, Rodnight R (eds) Research methods in neurochemistry, vol 5. New York, Plenum, pp 91–112
46. Pelligrino DA, Musch TI, Dempsey JA (1981) Interregional differences in brain intracellular pH and water compartmentation during acute normoxic and hypoxic hypocapnia in the anesthetized dog. Brain Res 214:387–404
47. Phelps ME, Barrio JR, Huang SC, Keen R, MacDonald NS, Mazziotta JC, Smith C, Sokoloff L (1982) The measurement of local cerebral protein synthesis in man with positron computed tomography (PCT) and C-11 leucine. J Nucl Med 23:6
48. Phelps ME, Barrio JR, Huang SC, Keen RE, Chugani H, Mazziotta JC (1985) Measurement of cerebral protein synthesis in man with positron computerized tomography: model, assumption, and preliminary results. In: Greitz T, et al. (eds.) The metabolism of the human brain studied with positron emission tomography. Raven. New York, pp 215–232
49. Raichle ME, Grubb RL, Higgins CS (1979) Measurement of brain tissue carbon dioxide content in vivo by emission tomography. Brain Res 166:413–417
50. Robertson JH, Wheatley DN (1979) Pools and protein synthesis in mammalian cells. Biochem J 178:688–709
51. Roos A (1965) Intracellular pH and intracellular buffering power of the cat brain. Am J Physiol 209:1233–1246
52. Roos A, Boron WF (1981) Intracellular pH. Physiol Rev 61:296–434
53. Roos A, Keifer DW (1982) Estimation of intracellular pH from distribution of weak electrolytes. In: Nuticelli R, Deamer DW (eds) Intracellular pH: its measurement, regulation and utilization in cellular functions. Liss, New York, pp 55–59
54. Rottenberg DA, Ginos JZ, Kearfott KJ, et al. (1984) In vivo measurement of regional tissue pH using positron emission tomography. Ann Neurol [Suppl] 15:98–102
55. Sacks W, Sacks S, Brebbia DR, Fleischer A (1982) Cerebral uptake of amino acids in human subjects and rhesus monkeys in vivo. J Neurosci Res 7: 431–436
56. Sbarra AJ, Karnovsky ML (1959) The biochemical basis of phagocytosis: I. Metabolic changes during the ingestion of particles by polymorphonuclear leukocytes. J Biol Chem 234:1355–1362

57. Siesjö BK (1978) Brain energy metabolism. Wiley, Chichester
58. Smith CB, Davidsen L, Deibler G, Patlak C, Pettigrew K, Sokoloff L (1980) A method for the determination of local rates of protein synthesis in brain. Trans Am Soc Neurochem 11:94/58
59. Smith QR, Pershing LK, Johanson CE (1981) A comparative analysis of extracellular fluid volume of several tissues as determined by six different markers. Life Sci 29: 449–456
60. Sokoloff L, Reivich M, Kennedy C, Des Rosiers MH, Patlak CS, Pettigrew KD, Sakurada O, Shinohara M (1977) The (^{14}C) deoxyglucose method for the measurement of local cerebral glucose utilization: theory, procedure, and normal values in the conscious and anesthetized albino rat. J Neurochem 28:897–916
61. Spector R, Coakley G, Blakely R (1980) Methionine recycling in brain: a role for folates and vitamin B-12. J Neurochem 34:132–137
62. Staffurth JS, Birchall I (1960) The measurement of the extracellular fluid volume with radioactive bromine. Clin Sci 19:45–53
63. Stossel TP (1974) Phagocytosis. New Engl J Med 290:717–723
64. Suleiman A, Spector R (1980) Methionine synthetase in mammalian brain: function, development and distribution. Life Sci 27:2427–2432
65. Syrota A, Castaing M, Rougemont D, et al. (1983) Tissue acid-base balance and oxygen metabolism in human cerebral infarction studied with positron emission tomography. Ann Neurol 14:419–428
65a. Syrota A, Castaing M, Rougemont D, Berridge M, Maziére B, Baron JC, Bousser MG, Pocidalo JJ (1985) Regional tissue pH and oxygen metabolism in human cerebral infarction studied with positron emission tomography. In: Greitz T, Ingvar DH, Widén L (eds) The metabolism of the human brain studied with positron emission tomography. Raven, New York, pp 285–303
66. Thulborn KR, DuBoulay GH, Duchen LW, et al. (1982) A ^{31}P nuclear magnetic resonance in vivo study of cerebral ischaemia in the gerbil. J Cereb Blood Flow Metab 2:299–306
67. Trivedi B, Danforth WH (1966) Effect of pH on the kinetics of frog muscle phosphofructokinase. J Biol Chem 241:4110–4114
68. Ui M (1966) A role of phosphofructokinase in pH-dependent regulation of glycolysis. Biochem Biophys Acta 124:310–322
69. Waddell WJ, Bates RG (1969) Intracellular pH. Physiol Rev 49:285–329
70. Waltz AG (1979) Comparative pathophysiology of ischemic stroke models: an evaluation. In: Price TR, Nelson E (eds) Cerebrovascular diseases. Eleventh Princeton Conference. Raven, New York, pp 11–17
71. Wheatley DN (1982) On the problem of linear incorporation of amino acids into cell protein. Experientia (Basel) 38:818–820
72. Wise RJS, Rhodes CG, Gibbs JM, et al. (1983) Disturbance of oxidative metabolism of glucose in recent human cerebral infarcts. Ann Neurol 14:627–637
73. Yamaguchi T, Waltz AG, Okazaki H (1971) Hyperemia and ischemia in experimental cerebral infarction: correlation of histopathology and regional blood flow. Neurology 21:505–578

Subject Index

GPSR Compliance

The European Union's (EU) General Product Safety Regulation (GPSR) is a set of rules that requires consumer products to be safe and our obligations to ensure this.

If you have any concerns about our products, you can contact us on ProductSafety@springernature.com

In case Publisher is established outside the EU, the EU authorized representative is:

Springer Nature Customer Service Center GmbH
Europaplatz 3
69115 Heidelberg, Germany

Batch number: 10704701

Printed by Printforce, the Netherlands